LIGNITE TECHNOLOGY

LIGNITE TECHNOLOGY

Edited by Perry Nowacki

NOYES DATA CORPORATION
Park Ridge, New Jersey, U.S.A.
1980

Library of Congress Catalog Card Number: 79-26051
ISBN: 0-8155-0783-6
Printed in the United States

Published in the United States of America by
Noyes Data Corporation
Noyes Building, Park Ridge, New Jersey 07656

Library of Congress Cataloging in Publication Data

Nowacki, Perry.
Lignite technology.

(Chemical technology review ; no. 146) (Energy technology review ; no. 53)
Includes bibliographical references and index.
1. Lignite. I. Title. II. Series.
III. Series: Energy technology review ; no. 53.
TP329.N68 662'.6222 79-26051
ISBN 0-8155-0783-6

FOREWORD

In this book an attempt is made to describe accurately the present technology used in the mining and combusting of lignite (including gasification and liquefaction) with the purpose of making this abundant domestic energy source more and more available for the generation of electric power.

Because the information in this book is taken from multiple sources, it is possible that certain portions of this book may disagree or conflict with other parts of the book. This is especially true of monetary values and opinions of future potential. We chose to include these different points of view, however, in order to make the book more valuable to the reader.

Cost figures provided are those given in the report cited, the date of which is always given. When the dates of the cost figures themselves are given, we have included them.

Advanced composition and production methods developed by Noyes Data are employed to bring these durably bound books to you in a minimum of time. Special techniques are used to close the gap between "manuscript" and "completed book." Industrial technology is progressing so rapidly that time-honored, conventional typesetting, binding and shipping methods are no longer suitable. We have bypassed the delays in the conventional book publishing cycle and provide the user with an effective and convenient means of reviewing up-to-date information in depth.

The expanded table of contents is organized in such a way as to serve as a subject index and provides easy access to the information contained in this book which is based on various studies produced by and for diverse governmental agencies, notably the Bureau of Mines and the Department of Energy, under grants and contracts. These primary sources are listed at the beginning of each chapter. The titles of additional publications pertaining to topics in this book are to be found at the end of each chapter.

Some of the illustrations in this book may be less clear than could be desired; however, they are reproduced from the best material available to us.

CONTENTS AND SUBJECT INDEX

INTRODUCTORY OVERVIEW

Lignite and subbituminous coal represent a large, relatively unused source of low-cost energy. Total coal resources in the U.S.A. have been estimated at 1,730 billion tons, and over 50% of the total is low rank, i.e., lignite and subbituminous. The tonnage of low rank coal that can be mined economically with the present technology is estimated to be about 190 billion tons. The ratio of "overburden volume to lignite volume" is most favorable when compared with that of hard coal. The deposits are located largely in the Northern Great Plains and Rocky Mountain regions, and bed thicknesses range from five feet up to more than 100 feet. Over 90% of this soft coal has less than 1% sulfur.

The current energy situation and the need to control sulfur dioxide emissions have greatly increased the present and planned use of lignite. Present production of low rank coals in the U.S. is about 75 million tons per year, nearly all of which is used for the generation of electric power. By 1985, greater use for power production and the development of plants for conversion to synthetic fuels from lignite are expected to increase the consumption of lignite to at least 300 million tons per year.

The profitability of "opencast" lignite mines has been amply demonstrated in the Rhineland district of West Germany which contains a deposit of about 60 billion tons of brown coal. Surface mines are easily recultivated and should not be compared with strip-mined fields.

Lignite displays unique properties compared to the higher rank coals that have traditionally been used in this country. Its high reactivity—evidenced by spontaneous combustion of many of the near-service reserves—poses transportation and storage problems, particularly for dried lignite. On the other hand, this high reactivity, plus the facts that lignite does not swell or cake upon heating, has high permeability, and produces ash with excellent sulfur-adsorbent qualities, make lignite a superior feedstock for gasification, liquefaction, and fluidized bed combustion.

This book discusses the present and future use of lignite for electric power generation and the problems associated with the use of low rank coals. Ash fouling of boiler tubes is the most serious operating problem. Research and commercial experience of utility companies burning high-fouling lignites from North Dakota and Texas have shown boiler tube fouling to be directly related to the sodium content of the fly ash. Many of the U.S. low rank coals are relatively high in sodium and produce a highly alkaline fly ash. Engineeringwise these problems can be attacked by the use of noncorroding alloys in boiler design, the installation of soot blowers, electrostatic precipitators and wet scrubbers.

The ash produced by lignite and related western soft coals is a basic ash in the chemical sense, meaning that it is an alkaline ash in which the sum of the CaO and MgO contents is greater than its Fe_2O_3 content. This means that it can be used for the chemisorption of SO_2 which is acidic, and should be well suited for mixing with clay in the making of fire bricks.

Yet the future of lignite utilization may not be in direct combustion, but in gasification and liquefaction. For this reason explicit details are devoted here to these two subjects and the solution of the engineering problems that they entail.

LIGNITE RESOURCES

The material in this chapter has been based on:

"Technology and Use of Low-Rank Coals in the U.S.A.," by G.H. Gronhovd and E.A. Sondreal of the Grand Forks Energy Research Center of the Energy Research and Development Administration, Grand Forks, ND, prepared for the Seminar on Technologies for the Utilization of Low Calorie Fuels held in Varna, Bulgaria, April 20-22, 1976.

DEFINITION OF LOW-RANK COALS

The accepted system for classification of United States coals by rank is that defined by the American Society for Testing and Materials (1). In accordance with this system, low-rank coals are those having a higher (sometimes called "gross") heating value (HHV) of less than 11,500 Btu/lb (6,390 kcal/kg), on a moist, mineral-matter-free (MMMF) basis, where moist means containing bed moisture. This definition separates lower-rank from higher-rank coals at a slightly higher level than that which is designated by the International Coal Classification System (2)—5,700 kcal/kg, or 10,260 Btu/lb, moist, ash-free basis—to separate "hard coals" from "brown coals and lignites."

Hence the U.S. low-rank coals that will be discussed in this chapter are on average slightly higher in heating value than the International System's brown coals and lignites, and correspond closely to the International System's Classes 9 through 13. The United States does not have coals of commercial interest in Classes 14 and 15.

The coals are completely nonagglomerating. Although the ASTM system provides a separate, lower classification called lignite B with less than 6,300 Btu/lb, MMMF basis, no commercial deposit with heating value this low is known to exist in the United States.

United States low-rank coals are further classified as follows:

Coal Rank	Higher Heating Value,* Less Than	Higher Heating Value,* More Than
Subbituminous	11,500 Btu/lb (6,390 kcal/kg)	8,300 Btu/lb (4,610 kcal/kg)
Lignite A	8,300 Btu/lb (4,610 kcal/kg)	6,300 Btu/lb (3,500 kcal/kg)

*Moist, mineral-matter-free basis.

The distinctions in HHV between low-rank and high-rank American coals, and between subbituminous coal and lignite, are shown in Figure 2.1 (upper portion).

Figure 2.1: Ranges of Higher (Gross) Heating Value and Proximate Analysis for U.S. Low-Rank and High-Rank Coals

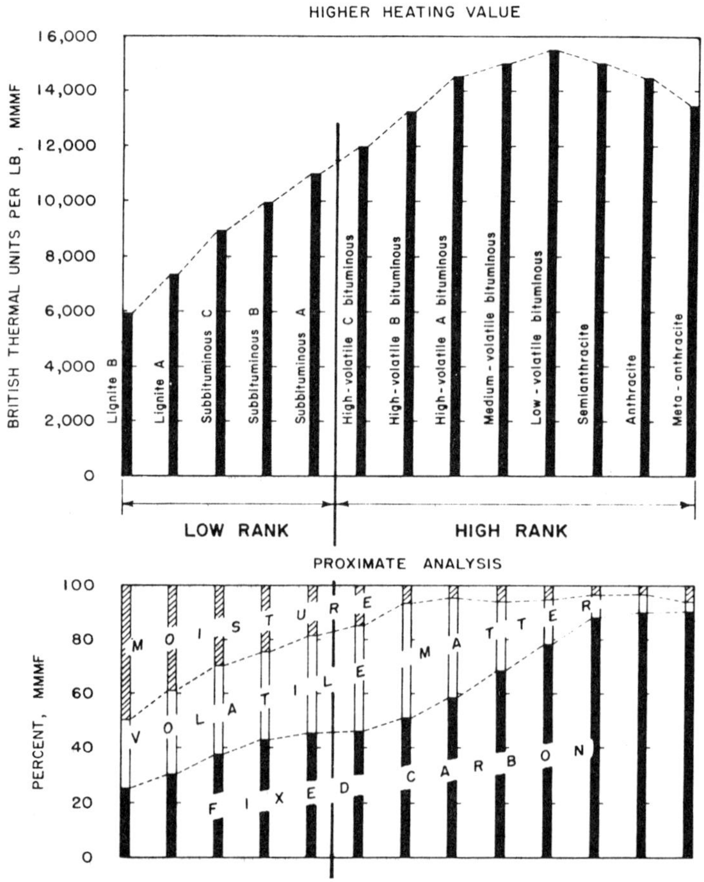

Source: "Technology and Use of Low-Rank Coals in the U.S.A.," April 1976.

The lower portion of Figure 2.1 illustrates characteristic differences in proximate analysis among the ranks of U.S. coals, all on the MMMF basis. In a general sense, subbituminous coals range in moisture content from 16 to about 32%, while lignite ranges from 36 to nearly 50% (MMMF basis). For both coal ranks, the organic portions are generally about equally divided between volatile matter and fixed carbon.

Ash contents for lignite and subbituminous coals are variable, but they tend to be lower than those of the high-rank coals. 15% ash content (as-mined basis) is unusually high, and there are major deposits of low-rank coal with less than 4% ash. A common ash content for United States low-rank coals now being commercially mined is 4 to 8% on an as-mined basis.

RESOURCES AND RESERVES OF U.S. LOW-RANK COALS

In official discussions and publications relating to the size of U.S.A. coal deposits, two basic terms are now used—"identified resources" and "demonstrated reserve base." These two basic terms can be broadly defined as follows (3):

Identified Resources: Quantity of coal of specified minimum thickness (14 inches for high-rank coal; 30 inches for low-rank coal) that is measured, clearly indicated, or geologically inferred in reasonably well mapped and explored coal basins, under a maximum overburden of 3,000 ft. Estimated quantities in unmapped and unexplored parts of known coal basins, or outside known coal basins, are not included.

Demonstrated Reserve Base: Quantity of coal of specified minimum thickness (28 inches for high-rank coal; 60 inches for low-rank coal) that is measured or clearly indicated, under a maximum overburden of 1,000 ft, and deemed to be economically and legally available for mining at the time of the determination.

Other terms are sometimes used for special purposes, including terms for estimates of recoverable coal.

The identified coal resources of the U.S.A., remaining in place as of January 1, 1974, as estimated by the Geological Survey (3), are shown below:

Coal Rank	Billions of Short Tons	Percent of Total
Anthracite	20	1.2
Bituminous	747	43.1
Subbituminous	486	28.1
Lignite	478	27.6
Total	1,731	100.0

The low-rank coals (subbituminous and lignite) constitute 964 billion short tons or 56% of the total identified coal resources of the U.S.A. On the basis of heating value rather than weight, using accepted average heating values (3), the low-rank coals make up 44% of the total. The demonstrated coal reserve base of the U.S.A. as of January 1, 1974, is estimated by the U.S. Bureau of Mines (4) as follows.

Coal Rank	Billions of Short Tons	Percent of Total
Anthracite	7	1.5
Bituminous	233	53.9
Subbituminous	165	38.2
Lignite	28	6.4
Total	433	100.0

The low-rank coals constitute 193 billion short tons or about 44% of the total demonstrated coal reserve base of the U.S.A. On a heating value basis, the low-rank coals make up 36% of the total.

In the table below the percentages of the demonstrated reserve base for subbituminous coal and lignite which fall within various sulfur ranges are shown (5).

	.... Sulfur Content, percent dry basis			
Coal Rank	1.0 or Less	1.1-1.8	1.8-3.0	Over 3.0
Subbituminous	89.5	8.1	2.0	0.4
Lignite	40.7	33.4	23.4	2.5

When these percentages are converted to the as-mined basis, about 90% of the total demonstrated reserve base of low-rank U.S. coal is shown to have sulfur contents of 1% or less.

Most (76%) of the country's identified resources of low-rank coals are located in the region designated as the Northern Great Plains Coal Province (Figures 2.2 and 2.3). Smaller deposits of considerable local importance exist in the Rocky Mountain Coal Province, as well as the Pacific and the Gulf Coast Provinces (Figures 2.3 and 2.4). Large, little-explored deposits have been found in Alaska.

Figure 2.2: Northern Great Plains Coal Province

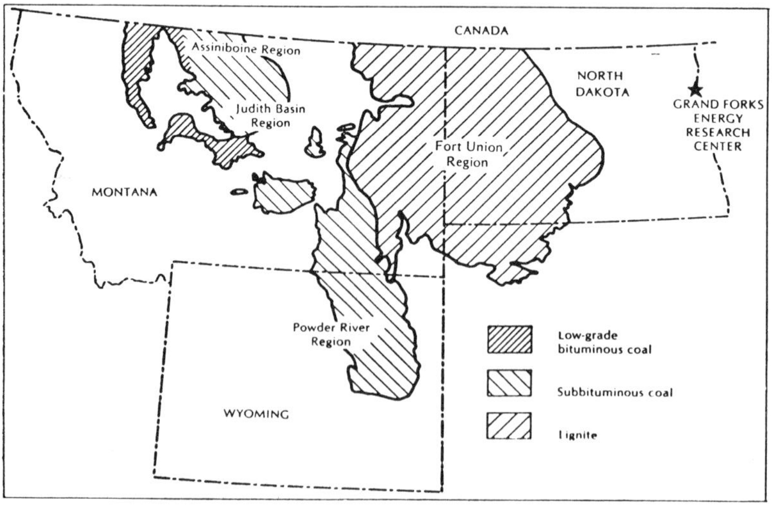

Source: Grand Forks Energy Research Center, 1978 Program Report

Figure 2.3: Coal Fields of the United States

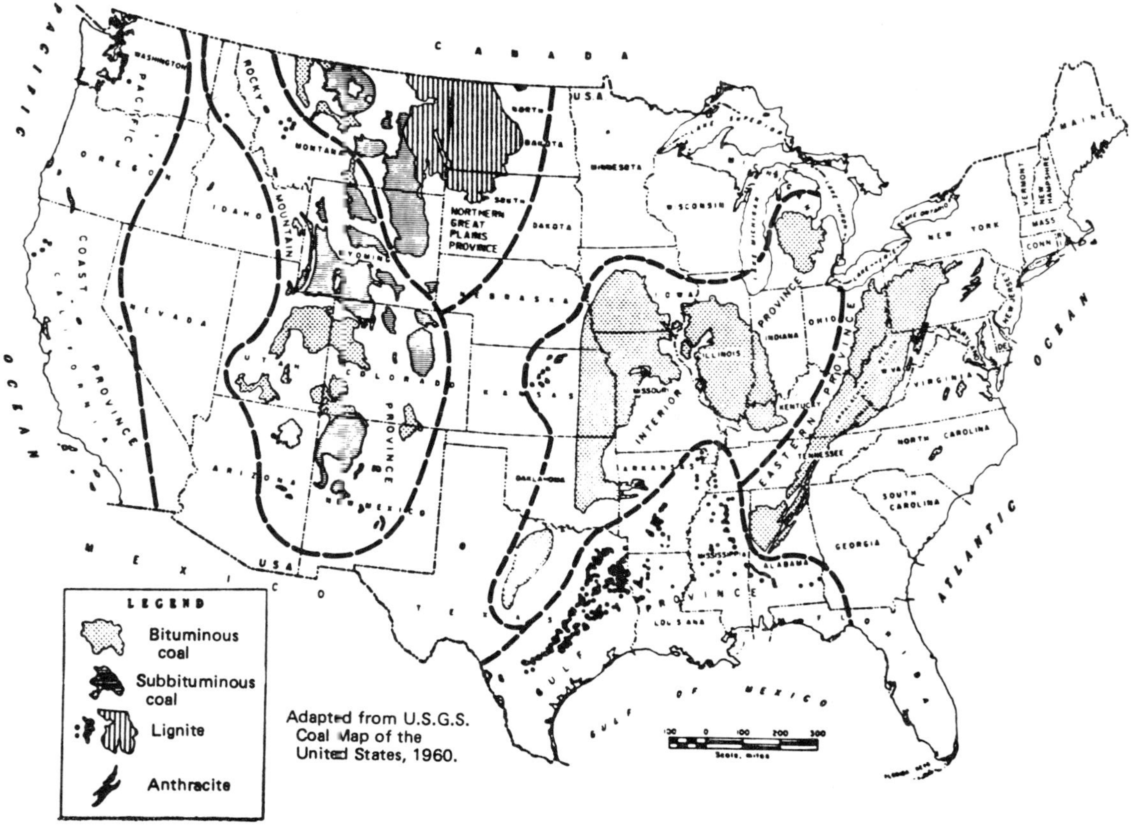

Source: "Technology and Use of Low-Rank Coals in the U.S.A.," April, 1976.

Figure 2.4: Coal Resources of Texas

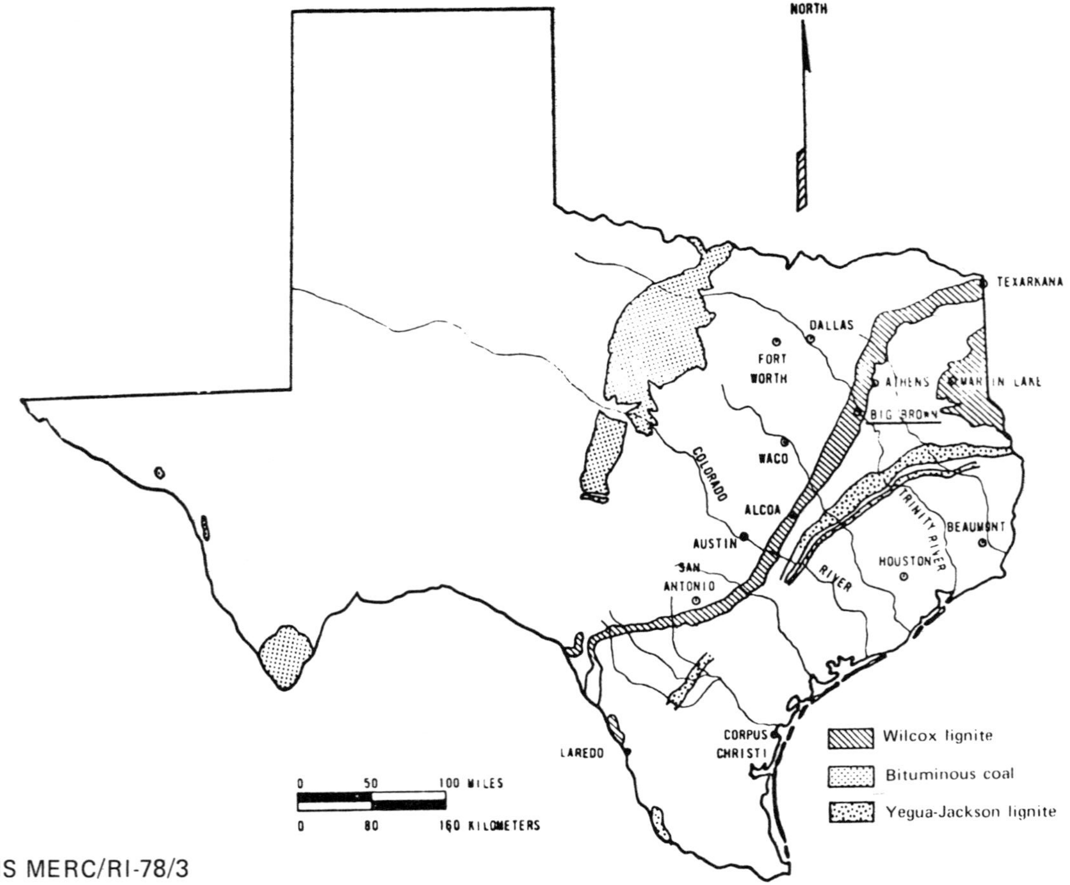

Source: NTIS MERC/RI-78/3

ASH COMPOSITION

Although the ash contents of all U.S. coals differ widely in chemical composition (reported as oxides), characteristic differences are evident between the older high-rank coals and the younger low-rank coals. Decrease in rank is usually associated with higher proportions of CaO, MgO, Na_2O, and SO_3, with lower proportions of SiO_2 and Al_2O_3. These characteristic differences are evident in the following table, in which data for 472 ash analyses (260 of bituminous ash, 212 of lignite ash) are summarized. Most subbituminous coals have intermediate ash compositions, with a general tendency to resemble the lignite ashes more than the bituminous ashes.

Typical Limits of Ash Composition of United States Bituminous Coals and Lignites*

	Bituminous, percent (6)	North Dakota Lignite, percent (7)
Silica, SiO_2	20 to 60	11 to 28
Aluminum oxide, Al_2O_3	10 to 35	8 to 14
Ferric oxide, Fe_2O_3	5 to 35	2 to 16
Calcium oxide, CaO	1 to 20	18 to 31
Magnesium oxide, MgO	0.3 to 4	2 to 9
Sodium oxide, Na_2O	0.2 to 3	1.4 to 6.5
Potassium oxide, K_2O	0.2 to 4	0.2 to 0.6
Titanium oxide, TiO_2	0.5 to 2.5	0.2 to 0.6
Phosphorus pentoxide, P_2O_5	0.0 to 3	0.0 to 0.6
Sulfur trioxide, SO_3	0.1 to 12	12 to 27

*Ranges shown represent approximately 95% of the total variation in the data considered.

The range of fusibilities of ashes from U.S. lignites does not differ significantly from that for U.S. bituminous coals, as indicated by results of a study of the analyses of 338 lignite and 135 bituminous ashes (8):

Ash Type	Range of Ash Softening Temperatures: Low	Average	High
Lignite ashes, °F	1,970	2,322	2,730
(°C)	(1,077)	(1,272)	(1,500)
Bituminous ashes, °F	1,940	2,355	2,930
(°C)	(1,060)	(1,290)	(1,610)

However, the effect of changes in ash analysis on fusibility is greatly different. Lignite ash is high in the major basic constituents, calcium, magnesium, and sodium and relatively low in the acidic constituents silica and alumina. The fusion temperature of lignite ash is lowered by increasing silica content and is raised by increasing calcium and magnesium contents. Bituminous coal ash by contrast is high in the acid constituents, and the effect of changes due to variations in major individual constituents tends to be reversed. A regression equation on the individual oxide constituents in lignite ash expresses the ASTM softening temperature as follows (8):

$$ST_{°F} = 2{,}326 - 6.9SiO_2 + 0.1Al_2O_3 - 4.3Fe_2O_3 - 128TiO_2 + 3.5P_2O_5 + 8.5CaO + 14.9MgO - 8.7Na_2O + 80K_2O - 5.1SO_3$$

A similar equation for bituminous coal ash is:

$$ST°_F = 1{,}164 + 12.1SiO_2 + 18.8Al_2O_3 + 7.2Fe_2O_3 + 83TiO_2 + 2.0CaO - 11.6MgO - 13.7Na_2O - 22.3K_2O$$

These equations, because of their simple linear form, best illustrate the comparative influence of elemental constituents in lignite-type ash versus bituminous-type ash. Other equations, also developed from data obtained at Grand Forks Energy Research Center, include curvilinear effects which allow a more precise prediction of fusion temperature.

PRODUCTION

Production of United States low-rank coal, which averaged about 51 million short tons per year during the 1964-1974 decade, or about 9% of the country's total coal production (9), is increasing very rapidly. Over the two years, 1973 to 1975, there has been an increase of 24.9 million tons, or 48%, as compared with only a 5% increase in production of the high-rank coals during the same time. High-rank coal is still predominant, as indicated below:

	Production of United States . . Coal, millions of short tons (9, 10, 11) . .			
Year	Low-Rank	High-Rank	Total	Low-Rank Percentage
1973	52.2	546.4	598.6	8.7
1974	63.7	546.3	610.0	10.4
1975	77.1	574.3	651.4	11.8
Gain, 1973-75, %	48	5	9	

Low-rank coal is produced in 5 general areas, or "coal provinces," in the United States. Four of the provinces are delineated in Figure 2.3; the fifth province is the State of Alaska, not illustrated. Over 80% of the low-rank production comes from the neighboring Northern Great Plains and Rocky Mountain Coal Provinces, as shown in the following tabulation (preliminary data) (11):

	. . .Low-Rank Coal Production, 1975. . .	
Coal Province	Millions of Short Tons	Percent
Northern Great Plains	38.65	50.1
Rocky Mountain	23.85	30.9
Gulf	10.11	13.1
Pacific Coast	3.70	4.8
Alaska	0.79	1.0
Total	77.10	100.0

Responding to the current pressure for increased energy in the face of limited oil and gas reserves, industry is planning very large increases in its capacity to produce. In a recent survey (12), volunteer information was requested from United States coal mining companies on their plans for expansion and new mine development over the next few years. Responses indicate that plans are in motion to add about 440 million annual short tons of additional capacity by the end of 1980. It is recognized that plans such as these are subject to unforeseen delays of many kinds; furthermore, the United States coal demand in 1980 is projected by the Bureau of Mines to be only about 800 million tons, including export requirements (13). Hence, the full addition of 440 million tons of capacity will probably not take place. Nevertheless, the survey indicates that coal companies in the United States see a marked increase in coal demand in the near future and are aggressively planning for it.

Results of the survey, in abbreviated form, are shown below:

Survey of Reported Planned Expansions of Capacity by 1980 (millions of short tons) (12)

High-rank coal		230.4
Low-rank coal:		
Northern Great Plains Coal Province	162.2	
Rocky Mountain Coal Province	30.4	
Gulf Coal Province	15.9	
Pacific Coast Coal Province	1.0	
Alaska	0.0	209.5
Total		439.9

A strong shift to low-rank coals is evident.

(1) Although low-rank production is currently less than 12% of total national production, the reported planned expansion in the low-rank area is almost 50% of total expansion.

(2) Planned low-rank expansion, fully achieved, would quadruple present low-rank production.

(3) The preponderance of low-rank coal development planning is focused on the Northern Great Plains Coal Province (Figure 2.2).

REFERENCES

(1) *Standard Specification for Classification of Coals by Rank, Designation D388-66,* American Society for Testing and Materials, Philadelphia, PA.

(2) United Nations, E/ECE/247, *International Classification of Hard Coals by Type*, United Nations, Publ. No. 1957 II. E., Min. 20 (1956).

(3) Averitt, P., *Coal Resources of the United States, January 1, 1974,* Geological Survey Bulletin 1412, (1975).

(4) U.S. Bureau of Mines, *Demonstrated Coal Reserve Base of the United States,* Mineral Industry Surveys, (1974).

(5) Hamilton, P.A., White, D.H., Jr., and Matson, T.K., *The Reserve Base of U.S. Coals by Sulfur Content: Part 2, the Western States,* BuMines IC 8693, (1975).

(6) Selvig, W.A., and Gibson, F.H., *Analyses of Ash from United States Coals,* BuMines Bull. 567, (1956).

(7) Sondreal, E.A., Kube, W.R., and Elder, J.L., *Analysis of the Northern Great Plains Province Lignites and Their Ash: A Study of Variability,* BuMines RI 7158, (1968).

(8) Sondreal, E.A., and Ellman, R.C., *Fusibility of Ash and Its Correlation with Ash Composition,* ERDA GFERC/RI-75/1, (1975).

(9) Westerstrom, L., "Coal—Bituminous and Lignite," and Federoff, D.R., "Coal—Pennsylvania Anthracite," *Minerals Yearbook 1973, Volume I, Metals, Minerals, and Fuels,* (1975).

(10) U.S. Bureau of Mines, *Coal—Bituminous and Lignite in 1974,* Mineral Industry Surveys, (1976).

(11) Merritt, P.C., "Productivity and Politics–Key Words for Coal Men in 1975 and 1976," *Coal Age,* Vol. 81 (2), pp. 82-86 (February 1976).
(12) Nielson, G.F., "Coal-Mine Development Survey Shows 492.6 Million Tons of New Capacity by 1985," *Coal Age,* Vol. 81 (2), pp.106-118 (February 1976).
(13) Dupree, W.G., Jr., and Corsentino, J., *United States Energy Through the Year 2000 (Revised),* BuMines Publication, (1975).

LIGNITE MINING AND RECLAMATION

The material in this chapter has been based on:

"Technology and Use of Low-Rank Coals in the U.S.A.," by G.H. Gronhovd and E.A. Sondreal of the Grand Forks Energy Research Center of the Energy Research and Development Administration, prepared for the Seminar on Technologies for the Utilization of Low Calorie Fuels held in Varna, Bulgaria, April 20-22, 1976

Technology and Use of Lignite, Proceedings of the Bureau of Mines-University of North Dakota Symposium, Grand Forks, ND, April 27-28, 1967 (Bureau of Mines Information Circular IC 8376), compiled by W.R. Kube and J.L. Elder for the U.S. Department of the Interior Bureau of Mines, May 1978

Technology and Use of Lignite, Proceedings of the Bureau of Mines-University of North Dakota Symposium, Grand Forks, ND, May 1-2, 1969 (Bureau of Mines Information Circular IC 8471), compiled by J.L. Elder and W.R. Kube for the U.S. Department of the Interior Bureau of Mines, 1970

Technology and Use of Lignite, Proceedings of the Bureau of Mines-University of North Dakota Symposium, Grand Forks, ND, May 9-10, 1973 (Bureau of Mines Information Circular IC 8650), compiled by G.H. Gronhovd and W.R. Kube of the Grand Forks Energy Research Center for the U.S. Department of the Interior Bureau of Mines, 1974

Technology and Use of Lignite, Proceedings of a Symposium Sponsored by the U.S. Energy Research and Development Administration and the University of North Dakota, Grand Forks, ND, May 18-19, 1977 (NTIS GFERC/IC-77/1), compiled by G.H. Gronhovd and W.R. Kube, Cochairmen, for the U.S. Department of Energy, 1977.

INTRODUCTION

All present low-rank coal production in the United States is from surface mining. Typical operations include topsoil removal and storage, overburden removal, coal removal, overburden replacement, and land reclamation. Overburden removal is by either power shovel or walking dragline, with the latter being most common. Draglines having 360 ft booms and 90 yd^3 (68.8 m^3) bucket capacities are in operation on low-rank coals. In the 1960s a bucket-wheel excavator was operated for overburden removal at the Glenharold mine in North Dakota. Its use was discontinued because of rock lenses in the unconsolidated overburden and conveyor-belt problems resulting from the cold climate.

After overburden removal, the exposed coal bed is drilled and blasted and then loaded into trucks using either power shovels or front-end loaders. Shovel bucket capacities are typically 20 to 40 yd^3 (15.3 to 30.6 m^3) and the front-end loaders have capacities up to 15 yd^3 (11.5 m^3). Trucks having capacities as large as 225 short tons are in use at some mines.

The low-rank coals in the northern Great Plains and Powder River regions occur in relatively thick seams ranging from 5 ft to more than 100 ft and typically with overburden from 50 to 200 ft. Under these conditions the productivity is high and may reach 100 short tons per man-day.

Reclamation of mined land is currently given great emphasis and has come to be regarded as an integral part of the mining operation. Regulations on reclamation differ somewhat from state to state (1), but generally no surface mining of coal is permitted until a plan for the mining operation, including reclamation, has been submitted to a designated state authority and accepted.

Typically, an acceptable reclamation plan stipulates the planned post-mining use of the land and requires grading to achieve suitable topography for that use. Any acid-forming materials or wastes are normally required to be buried. Top soil, to the extent that it is available, may be required to be spread over the graded earth. Revegetation appropriate to the climate and the planned use is always required, and a period of land management may also be stipulated to insure the success of the new growth.

The climatic conditions and terrain under which American low-rank coal is mined vary widely, from areas of mild winters and humid summers to areas of harsh winters, searing summers, and low precipitation. Most low-rank coal is produced under the latter conditions, typical of the northern Great Plains and the Rocky Mountain coal provinces, where establishment of new growth is a challenge. Smaller coal tonnages are mined under the favorable conditions of the West Coast and the coast on the Gulf of Mexico.

OPERATIONAL ASPECTS OF LIGNITE MINING

Initial Stripping Operation

In many instances, locating the first strip pit on a lignite bed so that only commercial quality lignite is obtained can prove difficult. In the majority of instances, the lignite along the outcrop of the bed is of inferior quality as it contains high moisture and high ash, has a low heating value, and a low ash fusion temperature. Such low-quality lignite can be found for as much as half a mile from the outcrop towards the center of the deposit. It is necessary to delineate the line on the lignite bed where the first pit should be located so that only high-quality lignite is exposed when the overburden is removed.

Ordinarily, this is accomplished by first drilling test holes and determining the actual limits of the lignite bed. Additional test holes are then drilled from the outcrop towards the center of the bed to a point where analyses of the drill samples indicate the lignite is of a commercial quality. The point where high-quality lignite is found is designated as the commercial lignite boundary.

Good practice dictates that the commercial lignite boundary should be located in the center of the first pit laid out. Laying out the pit in this manner offers assurance that the highest fraction of the lignite stripped will be suitable for industrial use. Very close supervision of the stripping must be maintained to assure that good quality lignite is mined. Should any slack pockets or other signs of inferior lignite appear on the outcrop side of the pit, the pit can be narrowed and then widened again as the lignite turns to better grade. This procedure gives a rather ragged looking line on the outcrop side of the pit, but it helps assure the quality of the initial lignite produced.

Drainage Control

Another problem that can prove to be very expensive and sometimes difficult to solve is that of mine drainage. Most stripping in North Dakota follows the contour of the land and the overburden is relatively shallow at the outcrop with the elevation of the surface usually increasing toward the center of the bed. Pits are normally located on the outer perimeter of the lignite bed, and drainage from the higher interior elevations has a tendency to collect in the open cut. Water accumulation in the pit can consist of both surface run-off and ground water. Every effort should be made to channel all surface run-off around the stripping operation.

It is not uncommon in North Dakota to receive as much as 2" of rainfall in less than a hour. Depending on the size of the stripping operation, considerable water can accumulate if not channeled away from the strip pits. A heavy cloudburst pouring into a pit will carry considerable silt that is deposited on top of the lignite, and sometimes equipment in the pit can be flooded.

Silt deposited on top of the lignite is expensive and difficult to remove. If it is not removed by recleaning the top of exposed lignite, the ash content is increased. Water pouring into a pit and running over and through a lignite vein naturally will raise the moisture content of the lignite for the next several months, thus diminishing its quality. Ground water in all cases must be removed from the pit, and under no circumstances should it be allowed to accumulate at the loading face. It should be stressed that any lignite loaded where there is water standing in or at the face must of necessity carry a higher moisture content than it would if the pit were well drained or pumped.

At the Beulah mine, the practice is to remove all ground water accumulating in a pit; and in the case of surface run-off, if it cannot be channeled around the pit, it is channeled into one specific area in a pit where proper pumping stations can be maintained. By doing this, the water is confined to one small area, and is easier to remove.

Cleaning and Blasting Exposed Lignite

Past analyses have indicated that the ash content in the top 6" of a lignite seam has a tendency to be high. Usual practice is to remove at least part of this top layer by cleaning the bed with bulldozers and scrapers immediately after exposure by stripping. All material loosened by these operations is removed. In addition, the surface of the portion of the seam set aside for each day's loading is again cleaned through use of front-end loaders and self-loading scrapers. A few years ago, a practice was inaugurated of sweeping the surface with mechanical brooms just before drilling, blasting, and loading.

Practically all lignite operations in North Dakota are using nitrate prills for blasting. For many years prills were not sensitive enough to be detonated in a 2½" hole; however, developments a few years ago have made it possible to use this type of explosive in a small-diameter hole.

Effect of Winter Weather on Operations

Probably one of the most severe problems faced by the lignite industry is the fact that peak production occurs during the coldest winter months when temperatures can plunge to -40°F. Ordinary metals exhibit peculiar characteristics at these temperatures and have a tendency to become brittle. During extremely cold temperatures, large loading booms would sometimes snap; large track corduroy, which is the track carrier, would break as though it were a piece of china; iron and cast steel gears in a tipple would shatter and fly through the air much like shrapnel; and huge walking cam guides on a large walking dragline would snap and break even though these castings weighed several tons each.

All this occurred during periods when the temperature was -30°F or lower and usually accompanied by a wind of 20 to 30 mph. With the advent of new specification steels this condition has been alleviated to some extent. It has been necessary to take precautions such as placement of electric heating strips on all the walking devices of draglines and on all gear cases, either in outside machinery or in preparation plants.

It is the practice at the Beulah mine to maintain a man on duty to check equipment every hour on any night when the forecast is for -25°F or lower. His primary duties are to make sure that the heating devices on shovels and draglines are in proper working order and to start various pieces of machinery in the tipple each hour so that the lubricant in the bearings will not become stiff. Consequently, operation can start at once in the morning. When this procedure has not been followed, it has sometimes taken 2 or 3 hours to get all of the equipment and machinery working. It is felt that such a procedure is a good investment.

Many years ago a large number of employees were utilized at each lignite operation, and the protection of these employees against extreme cold was difficult; however, the number of employees needed at an operation has diminished drastically with the advent of large equipment, and the equipment and the machinery used have evolved so that most employees are enclosed in a heated cab or are in a heated area.

Haulage Roads and Bearing Surfaces

Anyone who has lived in the northern United States is familiar with frost boils which occur in some roads each spring where rather severe winters are experienced. These same frost boils can occur in a mine haulage road, and cause difficulty when moving trucks or equipment. The problem of haulage roads has not been great in the past because the largest trucks being used were of 35 ton capacity. Recently trucks of 100 ton capacity or more have been introduced and the problems associated with haulage roads will, undoubtedly, be magnified.

Good drainage must be provided on all haulageways. In order to assure that heavy snowdrifts will not accumulate on the roadways, the surface of the haulage roads should be constructed at least 2 ft above the surroundings. It is good practice to install snow fences wherever needed, and to wet the road surface on a continuous basis shortly before freeze-up occurs each year. Wetting allows the top 6 or 8" of the roadway to freeze, and protects against deeper penetration from frost. A hard surface is provided for winter use resulting in lowered road maintenance.

It is anticipated that with the advent of large haulage units, all haulage roads will be constructed with water compaction to standards equal to or above those maintained by the State Highway Department. It will be necessary to carefully select and grade all materials used in the road construction; and, because gravel is relatively scarce in western North Dakota, it must be imported in order to build a proper road surface. Road surfacing materials will have to be carefully selected, and the binder added will have to be carefully controlled to avoid too much plasticity in the road surface.

Because of the scarcity of gravel in lignite producing areas, a substance called scoria is often used. The name scoria was given to this material by the Lewis and Clark Expedition, who associated it with volcanic action; however, the material is nothing more than a fired clay that has resulted from a lignite bed catching on fire and baking the clay above it. Most of this lignite burning occurred in prehistoric times when lignite veins which outcropped to the surface were set on fire by natural causes. Scoria is very similar to the average building brick in hardness and structure. It is quite apparent that if large haulage units are to be used, better surfacing material must be found.

The bottom clay underlying lignite beds in some areas of North Dakota can often give rise to considerable concern. Often the bottom clay is not capable of supporting the ordinary loading shovel as the bearing pressure of the tracks exceeds the bearing strength of the clay. At the South Beulah mine it is necessary to use large flotation tracks on the loading shovel in order to prevent the shovel from sinking into the bottom. Probably, as a precaution, all loading shovels will be fitted with flotation tracks wherever possible in future operations at this location.

The low bearing strength of the bottom clay often will not support the loads imposed on it by the overburden piles resulting from stripping. When the lignite is removed from the pit, the clay on the bottom moves, causing the overburden to slide into the pit and sometimes trapping and covering the loading shovel. This condition is usually encountered when stripping is conducted in very deep overburden and can be prevented by rehandling some of the overburden to keep the height of the overburden piles lower than would occur in one-pass stripping.

Magnitude of Proven Reserves

Years ago, reserves of 10 or 15 million tons were considered adequate for a single mine, because there were no large powerplants. This is no longer the case, as there are plants, either in operation or to be constructed, which will consume from 1 to 2½ million tons of lignite each year. This means that a plant using 2½ million tons of lignite per year would need a reserve over a 30 year period of approximately 70 million tons. In order to have recoverable reserves of 70 million tons, it probably would be necessary to have total reserves of 90 to 100 million tons, because all lignite cannot be recovered because of various economic reasons such as excessive overburden.

It is difficult to find such large reserves that are readily accessible, near a railroad, and of commercial quality. Each mining operation has in the past few years launched rather extensive programs in an effort to enlarge known reserves to meet future requirements.

Sodium Oxide Content of Lignite Ash

For years it has been known that lignite at times caused fouling in a boiler. However, most boilers were small and loads were not demanding so fouling, although a nuisance, did not unduly jeopardize boiler operation. In 1959, the Otter Tail Power Co. put a pulverized-lignite fired plant into operation at the Hoot Lake Station in Fergus Falls, Minn. It soon became apparent that there were instances when the lignite shipped to this station caused extensive fouling, resulting in loss of production. Each time the boiler unit went down for cleaning, production at the mine was also curtailed.

Over a period of two years, it became apparent that there were some areas in the mine from which the lignite, when burned at the Hoot Lake Plant, seemed to have a much higher tendency to cause fouling than the lignite from other areas of the same mine. Fouling lignite could not be defined by routine analyses.

By 1960, areas in the mine which caused trouble had been defined from experience, and such areas were bypassed when plant loads were going to be high. During the summer months when loads were low, lignite from the suspect areas was mined. Under conditions of low load, no particular trouble was experienced with fouling. It became apparent that there were areas from which the lignite could not be utilized by the Hoot Lake Station during periods of high loading.

Research conducted jointly by the Bureau of Mines, the Otter Tail Power Co., and the Knife River Coal Mining Co. indicated that the Na_2O content of the lignite ash appeared to have a direct relationship to the amount of fouling that could be expected when the lignite was burned. Various boiler tests indicated that in order to avoid excessive fouling and subsequent shutdown of the boiler under high load conditions, a low Na_2O content had to be maintained in the lignite ash.

In order to determine the magnitude of the problem, test holes were drilled in lignite beds that had been stripped at both the North and South Beulah operations. These holes were auger-drilled on 150 ft spacing. Analyses of the ash were made including the Na_2O content. Involved was a block of approximately 900,000 tons of lignite for which the sodium content of the ash was known for every 150 ft along the stripping operation. When these results were plotted on mine maps, it was easy to see that the high sodium areas corresponded rather well to those areas where experience had indicated that the lignite caused extreme fouling in the boiler.

The best method found for control of the sodium content was to mix lignite from high sodium areas with lignite from areas showing a low sodium content. This mixing was accomplished by using two shovels at each loading operation and loading trucks in a definite system alternating high- and low-sodium content lignite. By this method, a rather rough blend of lignite was obtained at each mine for which the approximate sodium content of the ash was known and to some extent, controlled.

As each consignment was made, the average sodium content was noted on the shipping notice. Because the shipment was several days in transit, the Otter Tail Power Co. had the opportunity to adjust loads among various stations if the sodium content was relatively high and the boiler load could be limited.

It is contemplated that mechanical mixing will have to be utilized most of the time if extreme fouling and shutdown is to be avoided at the Otter Tail plant. Cars are shuffled in the yard to obtain a proper sodium oxide content delivered at the plant. Mechanical mixing at the mine is an extra expense; however, the ability to keep the plant on the line and to prevent shutdowns of the boiler that also cause loss of production at the mine offset the extra cost.

MINING OF TEXAS LIGNITE

Area Stripping

Texas lignite is surface mined using area stripping methods. Overburden is removed by dragline side-casting using machines with buckets of 0 to 01 yd^3 and booms of 175 to 335 ft (Table 3.1). Currently no blasting is required, but is planned for the San Miguel project in South Texas. Bucket size and boom length have historically increased. Dragline productivity ranges from 235,000 to over 300,000 bank yd^3 per year per yd^3 of bucket capacity. Overburden readily packs into the bucket and very high average bucket fillability is the rule (2). Sticky material causes intermittent problems in dumping the bucket clean on each pass.

Stripping starts with a box cut along the lignite outcrop line at the 15 to 25 ft cover-depth line (at shallower depths lignite is highly oxidized and poor grade). 15 ft is typical for East Texas and 25 ft for South Texas. Successive pits are cut downdip into deeper cover. Pit widths average 120 ft (Table 3.1), but may be 160 to 180 ft wide in shallow cover (2). The narrower 100 ft pit at Darco is a function of the smaller equipment used there. Pit lengths are 3,000 to 12,000 ft. Highwall slope is 60 to 70°.

Table 3.1: Mining Equipment and Mine-Mouth Transportation Methods

Mine	Draglines	Pit Length Pit Width (ft)	Loading* (yd^3)	Mine Mouth Transport: Trucks (ton)	Mine Mouth Transport: Other
Sandow	35 yd^3 220 ft boom 95 yd^3 360 ft boom**	3,000–8,000 120	8	45 100***	2 mile conveyor belt, 600 fpm
Big Brown	70 yd^3 285 ft boom	12,000 120	16	120	
Monticello	60 yd^3 260 ft boom	10,000 120	16	100	2,500 hp, 25 kV catenary, 60 Hz electric locomotive; 15 to 20 100-ton cars, Texas Utilities RR; 80 100-ton cars, Louisiana and Arkansas RR.
Martin Lake	94 yd^3 335 ft boom 94 yd^3†	10,000 120	16 16† 10††	100 100†	2,500 hp, 25 kV catenary, 60 Hz; electric locomotive; 15 to 20 100-ton cars, Texas Utilities RR.
Darco	9 yd^3 175 ft boom 25 yd^3 275 ft boom†	3,000–4,500 100	2 4 5†††	25	
San Miguel	30 yd^3 215 ft boom 64 yd^3 275 ft boom	3,000–10,000 120	No shovel front-end loaders or mining machine	150	

*Shovels unless indicated otherwise.
**In 1979 and 1980.
***In 1978.

†To be added.
††Backhoe.
†††Front-end loader.

Source: NTIS GFERC/IC-77/1

Stripping ratios are less than 10:1 (yd^3 overburden per ton of lignite mined). Seams greater than 3 ft thick are mined. Mining depths are usually less than 120 ft.

Lignite is loaded for direct transport to a nearby generating station. No blasting is required prior to loading, but will be done at the San Miguel mine. Shovels, supplemented with front-end loaders and backhoes, are used for loading, clean-up, and parting removal. Partings greater than 3" thick are removed by loading to the parting, bulldozing it against the spoil, and loading the remaining lignite on a second shovel pass.

Mine Mouth Transport

Lignite is transported by truck (25 to 150 ton), truck-rail, and truck-conveyor belt. At the Big Brown mine transport is exclusively by truck with one-way haul distances of up to 10 miles. Truck-rail is the combination in use at Monticello and Martin Lake. The railroads were constructed by the Texas Utilities Company. These lines are electrified and involve runs of 6 to 12 miles. Each train, pulled by a 2,500 hp, 25 kV, 60 Hz locomotive, consists of 15 to 20, 100 ton bottom-dump cars which average 5 round trips per 8 hour shift.

Lignite is transported by truck to the loading site, loaded from overhead silos, and dumped into underground storage at the plant. The train does not turn around when returning from the plant because the caboose is equipped with the same controls as the locomotive (3). Eliminated is the land requirement for turning and a half mile of extra track.

The Monticello plant besides being supplied lignite via rail from Winfield will also receive lignite via rail from Martin Springs approximately 35 miles to the west. The run from Martin Springs will be by unit train (80, 100 ton cars) delivered by the Louisiana and Arkansas Railroad to a Texas Utilities siding south of Monticello.

The Sandow mine at Alcoa, Texas, is the only mine using a conveyor belt. About 2 miles of belt is used to annually move 2.1 million tons of lignite; deeper cupping can increase capacity up to 5.6 million tons. Belt speed is 600 ft/min.

Future Trends

Mining will continue to be done by dragline side-casting. Machines with longer booms (360 ft) will be able to dig deeper and place spoil in one pass. Mining depth is not likely to exceed 150 ft. This depth limitation is imposed by boom length and hydrology. The Wilcox Group is an important aquifer in Texas and mining below the water table in this water-rich unit will be difficult. Water in the highwall will cause slope instability and in the floor vertical movement and stratal distortion.

Previous mining to 160 ft at Alcoa ended in failure. From a hydrologic standpoint, deeper mining may be possible in the Jackson Group and Yegua Formation since they are generally not aquifers and, relative to the Wilcox Group, are water poor and mud rich. Uranium surface mining in the Jackson of South Texas extends to 350 ft.

The Monticello operation signals a possible future trend to the mining of several deposits tied by rail to sustain large steam-electric plants. Mining will become more complex as thinner, more lenticular seams with attendant higher stripping ratios are mined. Ratios are expected to increase to 15:1 or possibly even 20:1. Lenticular seams present a problem in maintaining pit length. Lengths of 1 mile or more are considered optimum; lengths down to 0.5 mile can be sustained. To mine multiseam deposits of 3 or more seams, draglines will probably be used in tandem.

More than one machine may be assigned to a pit, a larger upper machine on the highwall side removing the overburden down to the first seam and another smaller machine on a berm on the spoil removing interseam material. Scheduling the machines becomes a problem and the size assigned becomes critical because the amount of material each machine is to move must be balanced so that one machine will not have to wait on the other.

Bucket wheel excavators (BWE) are being evaluated for use in Texas. Opinion is that BWE have utility in mining multiseam deposits. Some deposits in Texas have as many as 6 minable seams. Seams as thin as 20" can be selectively mined (4). Commonly a lateral block-cut method is used (5). Soft material without hard nodules or boulders is required. Texas overburden is unconsolidated and weak.

Rock mechanics data on overburden from the Alcoa mine are as follows: tensile strength, 23 psi; uniaxial compressive strength, 165 psi; and unconfined shear stress, 175 psi (6). The overburden exposed at operating mines is generally free of nodules. BWE merit serious consideration in high-volume applications where material must be mined rapidly and continuously and transported some distance from the pit.

Reclamation

Reclamation in Texas is mandated by state law, effective January 1, 1976, and administered by the Railroad Commission of Texas, Surface Mining and Reclamation Division (7). Prior to this time, Texas Utilities Co., Alcoa, and ICI United States, Inc. were voluntarily reclaiming mined land. Adequate rainfall (35 to 50"/yr) with reasonably good distribution, rolling topography, unconsolidated overburden, and moderate sulfur content combine to favor successful reclamation.

Flat to moderately rolling country is a topographic configuration easily reproduced by redistribution and remolding of parallel rows of unconsolidated spoil. Top soil is only slightly more fertile than the overburden; hence, it is not segregated in mining. Mining has increased the content of exchangeable bases in mined soils due to mixing and movement to the surface of cations which have been leached into lower horizons or were released from parent materials at lower soil depths. Furthermore, spoil has water and nutrient holding capabilities superior to unmined material (8).

Maintaining drainage waters at suitably high pH has not been a problem to date, since the lignite has a moderately low sulfur content and the overburden is low in sulfur. Some mine waters have shown high sulfate content (775 to 1795 mg/l) (9). The Texas Railroad Commission is requiring monitoring to determine the impact of mining on ground-water quality and quantity. No assessment can be made of the effect of disrupted aquifers, disturbed recharge areas, and recharged mine waters. Surface pH's of 3.8 to 6.0 have been recorded in the reclaimed spoil (8). Low spoil pH is controlled by liming; for example, 4 to 5 tons per acre at Alcoa.

The land use trend in the lignite belt has been toward improved pasture acreage. Reclamation emphasis south of the Trinity River in east-central Texas is on revegetation for pastureland or hay farming (10). At Big Brown forage grasses with rapid growth and spreading characteristics such as Coastal Bermuda grass are used to revegetate the spoil following mining. Legumes such as Uchi Arrowleaf clover are planted for winter cover and nitrogen fixation.

Spoil is deficient in organic matter, P, and N, and requires fertilization for good yield. Coastal Bermuda grass yielded 1,514 lb per acre of dry matter in the year of establishment fertilized with 300 lb per acre N as NH_4NO_3 and 240 lb per acre of P_2O_5. Arrowleaf clover produced 32,153 lb per acre of green matter with 100 lb per acre N as NH_4NO_3 and 200 lb per acre of P_2O_5.

With proper fertilization and management the yield of grasses, legumes, and grain sorghum appears excellent. Ultimately, the long term productivity of the land will be the criterion for successful reclamation. North of the Trinity River in East Texas, reforestation, consistent with the area's traditional land use, will play an important role in reclamation. Loblolly pine has been planted at Darco.

In South Texas reclamation will be more difficult as rainfall decreases to 20" per year southward. Revegetation may occur naturally on a time scale that is unacceptable to society (decades or even centuries). It may be impossible, because of insufficient rainfall and alkali soils, to establish any kind of vegetation on some mined land. Revegetation in South Texas can probably be achieved only with major sustained inputs of water, fertilizer, and management (11).

Reclamation cost is 5 to 10 cents per ton of lignite mined. This estimate is based on a cost of $500 to $1,000 per acre to reclaim, 1,490 tons extracted per acre-ft, and a 7.5 ft seam.

MINING OF ESTEVAN LIGNITE (CANADA)

Location and Geology

The Estevan lignite area is located in the southwestern portion of Saskatchewan, just north of the United States border and the state of North Dakota. Lignite has been mined since the turn of the 20th century with all production prior to 1930 being produced from underground operations.

The Truax-Traer Coal Company introduced strip mining to the area and by 1956 underground operations had ceased with all of the tonnage being produced using strip mining techniques. Over 2 million tons per year are produced by three companies: Manitoba & Saskatchewan Coal Company (Limited), Battle River Coal, and Utility Coals Ltd. Heretofore, over 100 mines were necessary to produce this amount.

The coal in the Estevan area, which is of lignitic rank, is a portion of the Ravenscrag formation of the Tertiary age. The Ravenscrag formation is equivalent to the Fort Union formation of North Dakota and consists of interbedded lignite seams in variable thicknesses of sand, clay, sandstone, and shale.

Generally speaking, the coal horizons are horizontal but have a slight dip to the south. In the west there are severe irregularities which have not been fully explained or detailed.

Four seams have been mined in the area, but correlation of the seams throughout the field has not been definitely established, particularly for the upper seams. The Department of Industry in Saskatchewan is working on a project to correlate the seams by a computer analysis of all drilling to date. Isopach maps were made of the individual horizons and seams with variable success. Because indications suggest that three seams are present in the field rather than four, the computer study is being continued with detailed analyses.

The Btu content of the lignite varies from east to west; the highest values are in the eastern part of the field. The range is from 6,000 Btu to 7,500 Btu. The moisture content is approximately 35%, with ash content from 5 to 7%.

As in other areas where electric power is generated from lignite, the problem of controlling the sodium content of the ash is of paramount importance. Tests recently concluded in the Estevan area indicate that 3" cores can be recovered with little or no loss of core. Isopach maps can be produced showing the sodium content, which could assist materially both mining and power companies.

Mining

Manitoba & Saskatchewan Coal Company (Limited): Two seams are mined by the Manitoba & Saskatchewan Coal Company. Assuming that four seams exist in the field, the company is mining the lower two seams because the upper two seams have been eroded in this area.

The lower of the two seams is the most extensively mined and averages 8 to 9 ft in thickness. The seam is quite regular in horizontal configuration and quality. A Marion 7400 dragline with a 200 ft boom and 10 or 11 yd^3 bucket does the stripping in a normal side-casting procedure. Pit width averages 80 to 100 ft with depth of overburden ranging from 60 to 80 ft. No overburden drilling or shooting is required except in some winter months when the frozen ground is shot to maintain the efficiency of the dragline.

The upper seam is stripped by a Marion 7200 with a 120 ft boom and 7 yd^3 bucket. Overburden depth averages 20 ft with the coal seam being 4 to 5 ft in thickness.

Only very small tonnages remain as a reserve of the upper seam, which is not as extensive in area as the lower seam. The lower seam has not been mined when below the upper seam because economics and sales have not warranted operating at the depths of the lower seam in these areas.

Coal is drilled, shot, and loaded primarily using an 8 yd^3 electric shovel. In addition a small 4 yd diesel shovel loads coal behind the 7200. Fifty ton haulers are used to haul the coal to the preparation plant. Standard methods of crushing, screening, and loading are used. Oil treatment is applied where necessary. Most tonnage is shipped by rail with some minor tonnages being sold locally in trucks. The slack is hauled in trucks under contract to the Estevan Plant of Saskatchewan Power Corporation. The company's 1,000 hp diesel locomotive switches in the mine yards and transfers loaded and empty cars to the railroad yards in Bienfait 5 miles away.

Battle River Coal: The seam mined is the lowest of the four seams. All other seams have been either eroded or mined previously. As at the Manitoba & Saskatchewan Coal Company, the seam averages 8 to 9 ft in thickness and is quite regular. Two draglines were used, a Marion 7800 and a Bucyrus Erie 500W. The 7800 was then moved cross-country to the Utility Coal Company operation. This involved walking the unit approximately 10 miles and required extensive planning because a main highway and the Souris River were crossed. Before moving, the 7800 uncovered a considerable tonnage to provide a reserve of stripped lignite.

The stripping is done solely with the remaining 500W, which is equipped with 12 to 15 yd^3 buckets and removes approximately 50 to 60 ft of overburden. The coal seam is usually not drilled or shot but is loaded directly by two 5 yd electric shovels. 40 and 50 ton haulers are used to haul the coal to the crushing and screening plant. Most tonnage is shipped by rail, although some minor truck tonnage is sold locally and slack coal is shipped to the Estevan Plant of Saskatchewan Power Corporation. As at the Manitoba & Saskatchewan Coal Company, switching of the railcars is done by locomotives owned and operated by the mining company.

Utility Coals Ltd.: This company has been and is supplying the coal required by the two generating plants of Saskatchewan Power Corporation, the Boundary Dam Station, and the Estevan Generating Station. Approximately 1.5 million tons per year are burned by these plants.

Drilling indicates the presence of two seams in the area although only one seam has been mined thus far. The correlation of the two seams with the seams mined by Battle River Coal and Manitoba & Saskatchewan Coal Company has not been definitely established. The seam being mined has rapid elevation changes and detailed planning is required for stripping.

Several pits are maintained to assist in balancing the sodium content of the ash. A 750B shovel, 15W dragline, and 7800 dragline do the required stripping. Overburden depths and pit widths vary. A 170B, 9 yd shovel, and a 100B, 5 yd unit are used for coal loading.

When the coal is deeply frozen, a dozer rips the coal to assist the loading shovels. 45 ton and 65 ton bottom-dump trucks haul coal to the Boundary Dam Station. Coal delivery to the Estevan Plant of Saskatchewan Power Corporation is contracted.

Table 3.2 summarizes (according to working function) the type and size of equipment used at the various mines.

Table 3.2: Major Equipment Utilized in Lignite Mining*

	Company		
Operation	M & S	Battle River	Utility Coal
Stripping	7400 dragline 7200 dragline	500W dragline	7800 dragline 750B shovel 15W dragline 5W dragline
Loading	120B shovel, 8 yd^3 54B shovel, 4 yd^3	120B shovel, 5 yd^3 (2)	170B shovel, 9 yd^3 100B shovel, 5 yd^3
Hauling	50 ton truck (3)	50 ton truck (5) 40 ton truck (3)	65 ton truck (3) 45 ton truck (6) 17 ton truck (contract haul) (9)

*Numbers in parentheses indicate number of units

Source: IC 8471

Char and Briquetting Plant

The concept of a briquetting plant in this area started with the Lignite Utilization Board of Canada, which was established jointly by the Federal Government of Canada and the Provinces of Manitoba and Saskatchewan in 1918. A Lurgi-type carbonizer was built shortly thereafter one-half mile from the present preparation plant of Manitoba & Saskatchewan Coal Company.

In 1927 the plant was operated by Western Dominion Coal Company; it was sold in 1936 to Dominion Briquettes and Chemicals Limited, and then sold to a division of Husky Oil Canada Limited. In late 1968, the plant was purchased by the Manitoba & Saskatchewan Coal Company.

Approximately 100,000 tons per year of lignite are used to produce between 40,000 to 45,000 tons of char with an analysis of 70 to 80% fixed carbon, 10 to 16% ash, and 6 to 12% volatile matter.

Of the char produced, about 15,000 tons are used to make a domestic fuel briquet for heating purposes with a heat content of approximately 12,000 Btu. The remainder is shipped as char to various industrial users. At present most of the char is used in the production of barbecue briquets.

A flow diagram gives the general schematic arrangement of the plant (Figure 3.1).

Marketing

Originally the principal lignite market was for heating both domestically and industrially with little lignite being used in the generation of power. However, it is in the area of power generation and industrial uses, other than heating, that the future of lignite will exist.

Figure 3.1: Flow Diagram for Char Plant, Bienfait, Saskatchewan, Canada

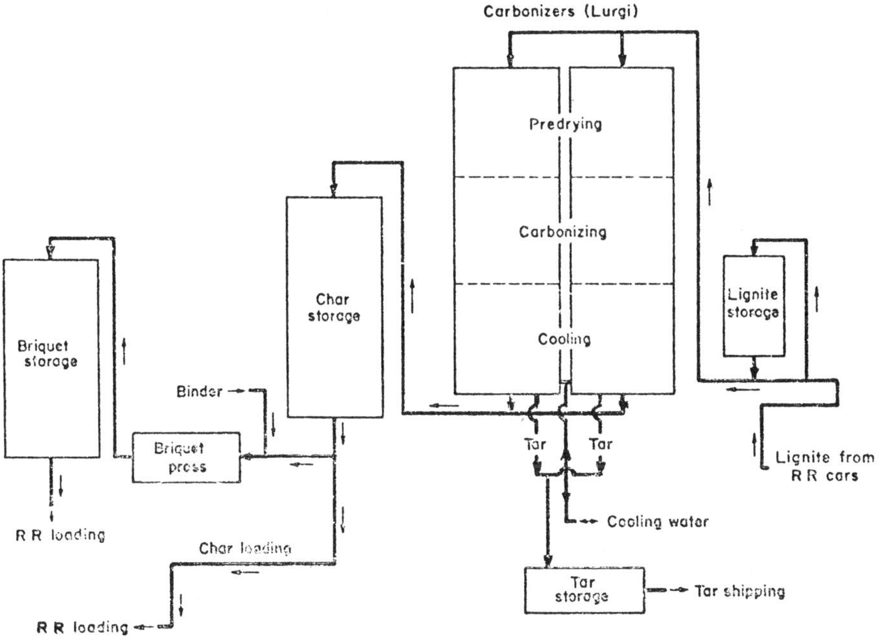

Source: IC 8471

Saskatchewan Power Corporation operates two plants in the area; the Estevan Generating Station and Boundary Dam Power Station, with coal supplied by Utility Coals Ltd. Small amounts of coal are supplied by the other two producers to the Estevan Generating Station.

The Estevan Station has a total 70 MW installed capacity, consisting of four units; 5, 10, 20, and 30 MW. Energy requirements from this plant are projected to require between 250,000 to 475,000 tons of lignite annually, although somewhat higher tonnages can be utilized depending on overall provincial power requirements.

The Boundary Dam Station is in the process of expanding. Two 66 MW generators are operating and another two 150 MW units will be completed in the near future. Additional units can be added if required and if economics warrant. The Station, with the total capacity of 432 MW, is projected to use 2.2 million tons of lignite annually.

Two other producers, Manitoba & Saskatchewan Coal Company and Battle River Coal Company, supply lignite to two generating plants of Manitoba Hydro at Brandon and Selkirk. The station at Brandon has a 120 MW capacity with an additional 100 MW unit under construction. The plant at Selkirk has a 156 MW installed capacity.

The growth requirements of Saskatchewan and Manitoba have made it desirable and economical to tie the two provinces together with an interconnection. This will provide better utilization of the existing and projected plants, whether hydro or thermal. The projected expansions and expected greater utilization of the thermal plants will provide for increased tonnages of lignite.

The domestic market for sized lignite is one that is steadily decreasing as consumers switch to other fuels or other methods for heating. The existing market for some tonnage in this broad area necessitates providing a well-prepared, sized product.

Industrial users, outside of power companies, purchase lignite for various applications. A good example of this application is the paper companies. Although, in general, this market has decreased, a substantial tonnage remains and will probably continue for some time.

As in the United States, research is continuing for the application of lignite in ore reduction, fertilizers, and hydrogenation. It is evident that increasing amounts of lignite will be used in power generation. The future markets for char or carbon derivatives will require additional investigation, and these are presently being carried on. Because the reserves of the Estevan field are sizable, the lignite from the area will continue to serve the Canadian and, possibly, the United States markets.

OPENCAST MINING TECHNOLOGY AT RHEINISCHE BRAUNKOHLENWERKE AG (WEST GERMANY)

Mining Practices and Utilization

The largest lignite deposit of Western Europe, which covers 970 square miles and contains 60 billion short tons of brown coal, is located in West Germany. The area forms a triangle with points at Cologne, Düsseldorf, and Aachen, as shown in Figure 3.2.

With modern technology about 13 billion short tons of lignite can be exploited to a depth of 1,650 ft in already developed and planned open mines. The seam averages 160 ft in thickness but occasionally branches out into as many as five separate seams.

During the last two decades mining technology was characterized by two developments: growing annual production rates, from 66 million to 105 million short tons during the period between 1950 and 1972; and the shifting from the conventional flat-open mine to modern deep-open mines.

The profitableness of the Rhenish lignite industry is attained by use of large operating units that can exploit the benefit of cost reduction most efficiently. Consequently, production has been concentrated in a few large mines. By 1960, all lignite corporations in the area merged into only one corporation, the Rheinische Braunkohlenwerke AG. During 1950 to 1972, the number of operating opencast mines was reduced from 23 to 6.

Figure 3.2: Map of the Rhenish Lignite District

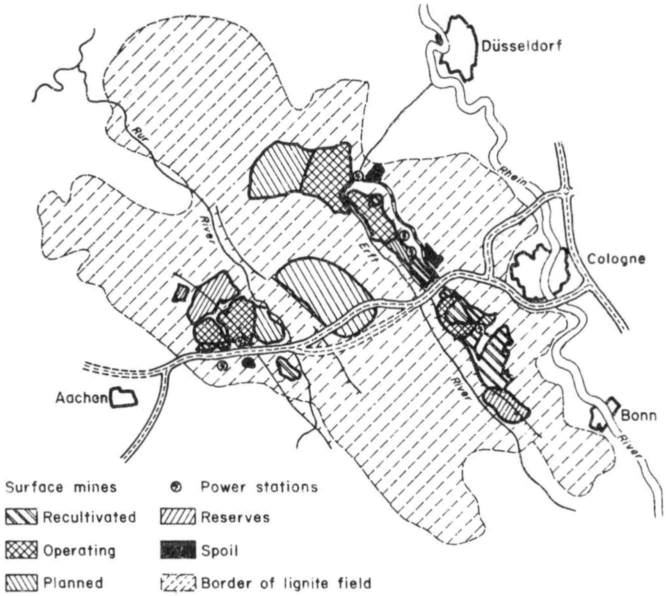

Source: IC 8650

The increasing depth of the mines is reflected by the growth of overburden excavation, 59 million yd^3 in 1950 compared with 280 million yd^3 in 1972. The ratio overburden volume to lignite volume, which indicates the profitableness of opencast lignite mines, soared from 0.8:1 in 1950 to 2.5:1 in 1972.

Realistic projects were developed for the opening of the Hambach mine. Five billion short tons of lignite reserves will be mined at a depth of 1,650 ft. The ratio of overburden volume to lignite volume will exceed 7.6:1.

For the time being, 80% of the lignite production, eventually reaching 125 million short tons annually, will be for thermal electric powerplants. The remaining 20% is for production of lignite briquettes.

In 1972, the district contributed 22% toward West Germany's total electric power production from thermal powerplants.

The large volume of excavated material to be moved, and the fact that mining depths are now about 660 ft and in the future will be as much as 1,000 ft, indicates that bulk transportation of material will be the crucial problem to be solved by the Rhenish lignite industry.

Under these circumstances, strip mining is not feasible, nor could belt bridges, spanning the distance between the excavation and the spoil bank levels, be used. The deposit requires opencast mines of large dimensions where the overburden is excavated on several levels by continuously operating devices and carried by high-capacity transport equipment along the mine border to the spoil bank levels in the mined-out sections of the open pit. A plan map of the arrangement for mining is presented in Figure 3.3.

Figure 3.3: Proposed Mining and Transportation Arrangement

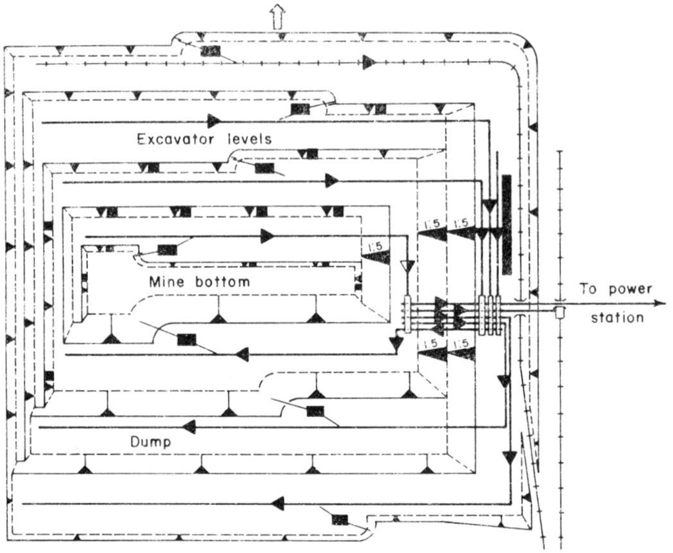

Source: IC 8650

In 1972, 42 million short tons of lignite and 97 million yd^3 of overburden was excavated in the Fortuna-Garsdorf mine, which is the largest open mine of the district. This equals 206 million short tons of material, which is the largest recorded annual output ever reached by any mine.

Bucket-wheel excavators with a daily output of 130,000 yd^3 of unconsolidated sediments are being used. Because the annual capacity of the Fortuna mine must be increased to produce 55 million short tons of lignite and 130 million yd^3 of overburden in the coming years, additional machines with doubled capacity are envisaged. These 260,000 yd^3 per day bucket-wheel excavators are expected to streamline operations substantially and increase the specific efficiency per man per shift from 100 to 130 short tons of lignite.

Design and dimensions of bucket-wheel excavators are decisively influenced by the desired excavation depth and characteristics of the soil. Stability of the slope depends on geologic strata. With increasing excavation height, the angle of slope must be reduced. This can only be attained through lengthening the bucket-wheel boom, which, however, tends to increase overproportionately both the total weight and the investment costs for the machine. An optimum relationship between capacity of the excavator and the height of excavation is necessary to keep the total excavation cost at a minimum. Another significant dimension is the bucket-wheel diameter, which is primarily gauged to the desired production capacity. Other important aspects are the ground stability of the various cut fractions and the demand for minimum specific transportation movements as related to the production volume.

It was deemed practical to adhere to the basic dimensions of the 130,000 yd^3 per day machine in development of the new bucket-wheel excavator. To guarantee unencumbered excavation and loading operations, the excavating step was isolated from the loading portion by means of a connecting telescoping belt bridge. The upward excavation height is 165 ft, and the maximum possible downward cut is 59 ft. The excavation range can be widened by lifting or lowering the excavator level above or below the level of the loading device, so that a total excavation height of 320 ft can be reached.

The machine measures 650 ft in length. A 70 ft diameter was selected for the bucket-wheel, which is equipped with 18 buckets holding 8.2 yd^3 each. With the 230 ft boom of the bucket-wheel, a block width of 330 ft can be reached. The excavator weighs 29 million pounds. The installed electric power will total 14,000 kW of which 2,500 kW is for the drive of the bucket-wheel (Figure 3.4).

Figure 3.4: Bucket-Wheel Excavator with Capacity of 260,000 yd^3/day

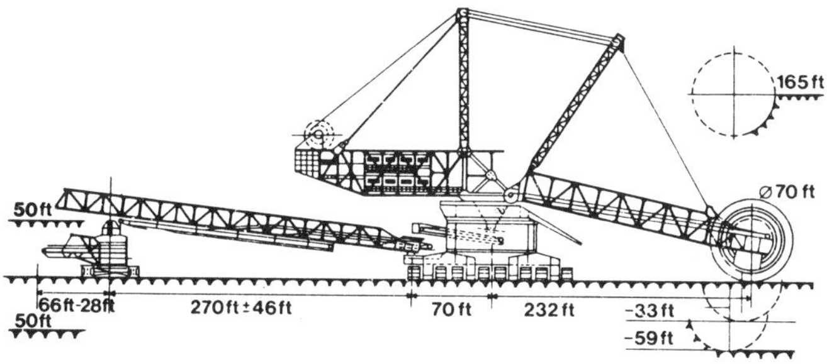

Source: IC 8650

For the development of the new machines, 2 decades' experience with the 130,000 yd^3 unit was available. Particular emphasis was given to maximum reliability of operation because shutdowns caused by operating disturbances are bound to affect the daily output of the mine proportionately to the increased capacity of individual machines. Consequently, the transportation system of the excavator was simplified, and the caterpillar drive was reinforced.

The doubled capacity of the excavator must be matched by corresponding developments at the stackers that dump the overburden. Capacity of the stackers will be increased by use of wider conveyor belts. The installed daily capacity of the stackers, 310,000 yd^3 per day, slightly exceeds capacity of the excavators so that temporary load peaks of the bucket-wheel can still be dumped by the stacker without necessitating shutdowns. A three-belt stacker with a receiving belt, supported on a separate driving unit, was chosen to keep the construction weight as low as possible. Accelerated speed of the unloading belt of the 330 ft boom also permits a further decrease of the construction weight of the stacker and widens the range of discharge (Figure 3.5).

Figure 3.5: Three-Belt Stacker Having a Daily Capacity of 310,000 yd^3/day

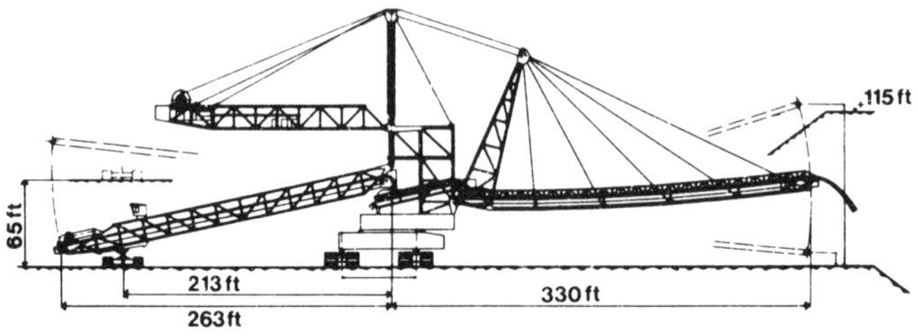

Source: IC 8650

The working range of a stacker is determined by boom length and gradient of the discharge belt; for faultless operation, the gradient should not exceed 12°. The new machine is able to stack up to a height of 115 ft. The depth in downward stacking is influenced by the fact that the discharged material must be built up on the bottom of the spoil banks to solidify the slopes. Assuming a 35° dumping slope, the resulting height will be about 200 ft. To widen this range, the machine is designed to enable the stacker to be placed up to 65 ft below the level of the dumpside belt. The stacker's unloading range thus totals about 380 ft in height.

With a view to the enormous capacity of the new machines, load concentrations in one spot on the ground must be avoided to enable the building of stable spoil banks. Therefore, the unloading techniques tend to prefer dynamic unloading operations that keep utilizing the full swinging of the unloading boom.

Transportation System

In the Rhenish district, railroad trains and conveyor belts are used almost exclusively for transportation. Over long distances and for small ascents, electric locomotives will continue to be used to haul the 125 yd^3 wagons containing overburden and the 149 yd^3 wagons containing lignite to the various coal consumers.

Within opencast mines, trains are usually not as applicable as conveyor belts because of their poor ability to climb. Also, in terms of streamlining operations, conveyor belts are superior to trains. All conveyor belts are operated and controlled at only one dispatching desk. Automatic devices control the unmanned driving and transfer stations and will stop the operations in case of emergency.

The most important device is installed in the loading funnels. A probe, based on electric capacity-monitoring, safeguards from overloading. Other essential control devices are being tried to check slipping of the belts on the driving drums, to monitor belt tensions, and to detect longitudinal cracks in the belts.

Automation keeps the man-hours for conveyor belt operations at a minimum. Similar reductions in railroad operations are limited because of the necessity for trackshifting.

For belt operation, the most economical system must be devised. Mine conditions are decisive. Size and shape of the minefield influence the length of the levels on which the volume transported depends. The dimension of the system also depends on the selected exploitation pattern. For so-called parallel operation, level belts lead to front-end belts, which connect the excavation levels with the spoil bank levels at the border of the opencast mine.

The material must be distributed to the different spoil bank levels to build up stability, and lignite must be kept separate from overburden; therefore, each level must be equipped at the junction points with transfer devices having back-and-forth sliding heads to permit loading on the different belts.

For the swinging operation pattern, a distribution point is established, if possible in the center of masses, around which the excavating and spoil bank levels are swung. The belt ways of the different levels meet at this point.

Both exploitation patterns are, however, not quite satisfactory in that the material is transported from the excavation point along the mine border to the spoil bank site.

A third possibility, the so-called diagonal belt conveyor pattern, was recently used for the first time in the Rhenish district. With this method, the excavation and spoil bank sides are connected diagonally, substantially shortening the distance for transport. The disadvantage is that all material to be transported passes to the bottom of the mine and is hoisted from there to the individual spoil banks or to the lignite hopper. The savings involved in reduced transport work will, however, offset this defect.

Conveyor Belts

The operation of the new types of excavators and stackers requires modern belt conveyor systems having twice the capacity of previous models. The belts are designed for an hourly capacity of 33,000 short tons. Driving stations can be equipped with up to 6 electromotors, each one with a capacity of 1,500 kW. The planned width of the wire-rope armed belts (manufacturing type St 5,400) is 118". The belt speed will be 21 fps.

One of the modern 1,500 kW driving stations is already operating at the Fortuna mine. Permanently installed compound-walking devices enable self-propelled motion of the 1.8 million pound station when the belt system is shifted.

Development of new machines and belt conveyors indicates that the Rhenish lignite mining industry has decided on capacity increases of opencast mining machinery to attain streamlining and cost reductions. Nevertheless, another increase in efficiency by raising the hourly capacity substantially beyond the 33,000 short ton mark will not be feasible unless transport technology takes a new course. In particular, new belt-driving devices are necessary. To meet this challenge, the Rhenish mining industry has joined forces with various scientific institutions and numerous companies of the component industries to find solutions to the problems.

Spoil Reclamation

In addition to the problems of opencast mining technology, several problems must be solved that stem from deep open-pit mining covering large areas. These could be summarized under the heading environmental impact. Especially two environmental factors influence the costs involved.

First, the Rhenish lignite district contains a population of 1 million at a density of 1,070 inhabitants per square mile, making it one of the most densely populated areas in Europe. The United States has about 47 inhabitants per square mile. Second, the lignite is covered by a layer of loess, which is one of the most fertile soils in the entire European community and is intensively cultivated. The major crops are grain and sugar beets, with some vegetables being grown in the immediate vicinity of major cities.

Large operating units with equally large production capacities are necessary to find and to finance technically feasible and economically reasonable solutions to the environmental problems that develop during, and particularly after, mining exploitation.

The mining areas include public railroads, highways, and even rivers. As mining operations expand, these have to make way, which means that houses, farms, and whole villages have to be moved. Since the beginning of the mining operations, 44 villages with 18,500 inhabitants have been resettled. An additional 9,000 people will have to be moved in the next two decades.

In connection with the resettlement of villages, the necessary technical and organizational measures are combined with efforts toward improved social conditions. Thus, the structure of the new villages and communities is considerably enhanced. Upon completion of the mining operations, the surface of the exhausted mines is restored so that the entire area can be used.

The major portion of the mining districts in the north was once mostly cultivated land. Accordingly, efforts are made to reclaim a matching portion of the spoil banks for agricultural purposes. To attain this goal, the loess, which may be as thick as 65 ft, is useful. The loess is excavated separately and a 3 to 7 ft layer is spread over the spoil bank surface either by dry spreading or by sluicing.

The south mining district was not cultivated when lignite mining began. Here, the high slopes of the spoil banks are reclaimed by forestation. The surface of the spoil banks is covered by a mixture of gravel and loess, and then planted with some 36 species of trees and 18 species of shrubbery. Lakes are created in the remaining unfilled hollows. The new landscape developed in the wake of lignite mining, lakes and woods in the southern district and farm lands interspersed with wooded slopes in the northern district, add valuable recreational areas benefiting the inhabitants of this densely populated country.

Only proper reclaiming will insure that further demands for land to be used for open lignite mining will be granted. Governmental authorities and autonomous corporations including representatives of the local population are watching the operations, and all must be satisfied. Moreover, it is imperative to familiarize the public with the exploitation measures and their economic significance.

RECLAMATION OF LANDS IN THE LIGNITE AREA OF THE NORTHERN PLAINS

Practices of Reclamation

Surface mining for lignite in the Northern Plains is expanding to meet current energy demands. The commitment to reclaim mined land to a high level of productivity has necessitated a great need for reclamation technology.

The lignite area of the Northern Plains is located in eastern Montana, western North Dakota, and northwestern South Dakota. The climate of the lignite area is semiarid, with annual precipitation averaging from 12 to 16", about half of which normally occurs during May and June.

Temperatures within the region exhibit great seasonal and diurnal fluctuations and range from -40°F in winter to 115°F in the summer. Average January and July temperatures are about 4° and 66°F, respectively, in the northern part and about 18° and 72°F, respectively, in the southern part of the region. The climate is adapted for cool-season grass and small grain production.

Agriculture comprises the primary land use of the area. A 1972 study (Table 3.3) of 20 counties included in the North Dakota lignite area showed that of the 18.8 million acres represented, 49% was cropland and 51% was native prairie. About 84% of the cropland acres was used for the production of small grains. Livestock production is the main industry in the prairie lands. Desired land use after mining will influence reclamation decisions regarding post-mining topography, species composition, and management. Regardless of land use, the reestablishment, growth, and production of vegetation is essential.

Table 3.3: Land Use (1972) in 20-County Lignite Coal Area of North Dakota*

Area	Acres (millions)	Cropland	Small Grainland	Prairie
		....	(%)	....
Northwest (6)	5.9	64	59	36
West central (5)	5.5	40	32	60
Southwest (7)	5.1	44	37	56
South central (2)	2.3	42	29	58
Total	18.8	49	41	51

*Data from ND Crop and Livestock Statistics Report No. 32, 1974. The numbers in parentheses indicate the numbers of counties in each region.

Source: NTIS GFERC/IC-77/1

In 1976, 10.3 million tons of lignite were strip mined in North Dakota, disturbing a section (640 acres) of land. Present projections indicate that lignite mining will increase substantially to meet energy requirements and to reduce dependence upon imported oil. Many federal, state, and private organizations are now involved in research to better understand, develop, and apply technology to both agricultural and non-agricultural facets of reclamation.

Early in the development of the modern reclamation effort, the need to integrate reclamation plans with the overall mining plans was recognized. By applying this principle, reclamation problems can be analyzed and mitigating measures planned for in the overall mining plan. Often, a mining operation requires several years for orderly initiation of its proposed mining and reclamation plan.

Since the late 1960s and early 1970s, a set of general mining and reclamation procedures have emerged and are being used in the lignite mining region. These procedures have evolved through research, legislation, and experience.

The first step is to obtain all the appropriate permits required by law. During this stage, information is gathered concerning the environment, lignite resource, land ownership, and characteristics of the soil and overburden for the proposed mining site. This information is evaluated by assembling into environmental impact statements, reclamation and mining plans. After permits are issued, the development of the mine headquarters and the mine itself begins.

In the actual mining process, soil materials suitable for plant growth (in North Dakota, up to 5 ft if available) are removed and stockpiled. This is usually accomplished with scrapers. The exposed overburden is then removed and side cast by dragline, exposing the various lignite veins. These are then loosened by blasting and loaded into large trucks by front end loader or shovels to be hauled to the tipple. The lignite is crushed and used to fire nearby mine-mouth power plants or it is shipped by rail to distant power plants.

After the lignite has been removed, the spoils are reshaped, usually by bulldozers or small draglines, and the stockpiled topsoil is then replaced by scraper over the smoothed spoils. The areas are tilled, fertilized, and seeded to grass and legume mixtures using conventional farming equipment and practices.

Problems Associated with Reclamation

One of the most severe concerns currently facing reclamation deals with differential subsidence and piping erosion. In general, this is caused by the slumping and settling of reshaped spoils and erosion along subsurface cracks which result in cavities beneath the spoil surface. These problems seem amplified when spoils are sodic and, thus, structurally unstable. Many questions concerning these phenomena and their prevention remain. Further research into these problems is currently being conducted.

Another area of concern is erosion control on newly spread surface soils. Because of the lack of vegetation to protect the surface, the reclaimed land is susceptible to wind and water erosion. Technology is available whereby this erosion can be reduced. Practices like mulching, shaping to less steep slopes, and prompt establishment of vegetation, can greatly reduce erosion losses.

Even though the replacement of suitable plant growth material over sodic and saline spoils has greatly reduced the problems associated with reclamation of these spoils, concern still exists for controlling salts. High sodium adsorbtion results in slow water infiltration and high salinity results in unfavorable water availability for plant use. Research is underway to better understand the movement of salt and water at the interface between saline or sodic spoils and replaced surface soil. These problems are especially important in areas where very limited amounts of suitable soil materials are available for stockpiling and replacement.

Postmining use will be substantially dependent upon reestablished vegetation, and prompt establishment of vegetation is the key to erosion control. Therefore, a more reliable procedure must be developed to establish vegetation. Currently, either one of two seeding seasons is commonly used for seeding perennial grasses; spring plantings (before June 1), and fall plantings (from late August until freeze). The success of either seeding period is determined primarily by weather conditions that follow seeding.

Another major concern of reclamation is long term sustained production of annual and perennial crops. Observations available at present are based on only a few growing seasons. The time factor in evaluating the success of reclamation has not been adequately determined. Research currently in progress is designed to monitor production of reclaimed land for several years over a variety of weather cycles. Only by such evaluation can long term productivity of reclaimed land be estimated.

Closely allied with sustained production is the determination of the best management practices for reclaimed land. This includes factors like soil fertility, cultural practices, livestock stocking rates and grazing period, natural succession of species, and species adaptability to reclaimed lands.

Reclamation Research by the Northern Great Plains Research Center

Reclamation research at the Northern Great Plains Research Center began in 1969. Early work was largely exploratory and helped to define basic problems. The characterization of the physical and chemical properties of overburden and spoils was the first approach taken in the development of the reclamation research program. Sandoval discussed several of these characteristics and how they relate to potential for reclamation (12).

A key factor was the amount of adsorbed or exchangeable sodium present in the overburden materials. This coupled with the frequent occurrence of fine-textured spoils of montmorillonitic clay mineralogy, resulted in growth media that severely restricted water movement and water availability for plant growth. The effect of these and other spoil characteristics on vegetation was discussed in more detail by Power (13).

From overburden and spoil chemical analyses, the common occurrence of sodium and magnesium sulfate salts in spoils of the Northern Plains has been documented. A study was, therefore, conducted to evaluate the response of eight perennial forage species to different spoil textures and contents of magnesium and sodium sulfate (14).

Observations were made under growth chamber conditions at germination, emergence-establishment, and vegetative growth stages of plant development. It was found that these species responded differently to similar salt levels and ionic concentrations and that sensitivity to salt levels and ionic concentration was not the same at all stages of plant development for a given species. Some species [fourwing saltbush (*Atriplex canescens*) and alkali sacaton (*Sporobolus airoides*)] showed great tolerance to salts during the vegetative growth stage, whereas Canby bluegrass (*Poa canbyi*) was sensitive to both increased salt concentration and to high magnesium ion concentrations.

Sandoval (12) also showed that the lack of plant-available phosphorus often limited plant growth in spoil materials. Power (15) reported that when Paleocene shale overburden was exposed to the atmosphere, nitrification of the exchangeable ammonium usually present on these shales would occur, resulting in increased nitrate nitrogen content for a short period after mining. Spoils generally contained no mineralizable nitrogen, however. These and other soil fertility problems are readily corrected by adding appropriate commercial fertilizers or other sources of plant-available nutrients.

Chemical amendments and mulches were added to sodic spoils to increase water infiltration and availability, limited by the sodic and fine textural conditions. Materials studied in 1970 included straw mulch, sulfur, calcium sulfate (gypsum), and topsoil. Best results were found where spoils were mulched with 2" of topsoil material. Grass yields were increased and exchangeable sodium decreased for the gypsum treatment, although time and natural precipitation were needed for the chemical reaction to occur (Table 3.4).

Table 3.4: Dry Matter Yield for Perennial Grasses Grown on Strip-Mine Spoils, 1972–1974.

	.Exchangeable Sodium in Spoils.		
			Average
	High	Moderate	All Tests
Treatment	....	(dry weight/lb/acre)	....
Check	<100 (5)	3,330 (3)	1,290 (8)
2" topsoil	1,550 (5)	4,270 (3)	2,570 (8)
Gypsum	380 (3)	3,730 (3)	2,060 (6)
Gypsum + 2" topsoil	1,220 (3)	4,270 (3)	2,720 (6)

Note: Numbers in ( ) represent number of experiment years included in mean.
Native prairie grasses on Grail silt loam at Mandan, 1972-74 = 1,320 lb/acre without fertilizer, 4,640 lb/acre with fertilizer.

Source: NTIS GFERC/IC-77/1

Later, Doering and Willis used calcium chloride, followed by leaching, to chemically improve a sodic spoil (16). This process is expensive and requires large quantities of water, but it does provide a technique for quicker reclamation under some circumstances.

The importance of characterizing overburden properties before mining encouraged a study of sampling methods. Power and Sandoval found that the drilling method affects the chemical composition (17). Methods that use a circulating fluid can potentially affect the composition of soluble constituents in the samples collected. Care must be taken to use a drilling method that will minimize alteration in sample composition.

Results of laboratory analysis of mine spoil indicated that some conventional laboratory procedures were more reliable than others. Also, data obtained by certain laboratory procedures are better correlated with potential plant growth than are other procedures. These procedures relate better to field conditions encountered. Sandoval and Power reviewed laboratory methods and recommended procedures for chemical analysis which they believe are most suitable for mined-land spoil and overburden in the western United States (18).

The ability of vegetation to become established and produce on reclaimed mined land is essential to most postmining land uses; recreation, livestock grazing, wildlife habitat, or crop production. Plant species have different requirements and adaptability for various environmental and cultural conditions.

A study is being made to evaluate the suitability of 36 grass and legume species for planting on reclaimed land. These species have been planted under three conditions: (1) on sodic spoil; (2) on sodic spoil covered with 6" of topsoil; and (3) on agricultural land at the Northern Great Plains Research Station.

Results so far show that the cool-season grasses [wheatgrasses (*Agropyron*), smooth bromegrass (*Bromus inermis*), and green needlegrass (*Stipa viridula*)] established and produced well on mined land (19). These are the species usually recommended for dryland seedings on nonmined land. Warm-season grass species, like sideoats grama (*Bouteloua curtipendula*) and blue grama (*Bouteloua gracilis*), did not become established nor produce well. Many questions remain unanswered about their requirements for establishment and growth on either nonmined or mined land.

Another species suitability trial was initiated in 1976, in which species were planted into varying depths of replaced subsoil and topsoil, to further determine specie specific replacement depths of surface soils. No data are available at this time from this study.

As the acreage of land disturbed by mining annually increases, the need for more reliable revegetation technology also increases. A primary factor in vegetation establishment each year is water. Techniques, which will lower salinity or sodium levels, will improve water movement and availability for plant establishment and growth. Water conservation techniques, like mulching, can also help increase water availability. But even with good water management, new seedlings are often subjected to severe water stress, which can result in stand failure because of limited and irregular precipitation patterns.

A current study investigates the effect of applying limited quantities of water to supplement low or irregular natural precipitation. Results to date show that applying limited quantities of water at certain critical times during the year of establishment can help insure successful stand establishment and can greatly influence species composition of the establishing plant community. With supplemental irrigation, seeding can be accomplished at almost any time during the growing season.

Successional sequences of species are also influenced by the time of seeding and amounts of water received. Research is also presently being conducted to determine if saline water, like that often pumped out of mine pits, can be used for supplemental irrigation.

The final test of reclamation will be the ability of the reclaimed land and its vegetation to withstand sustained use and occasionally misuse. Since about 51% of the lignite coal lands are currently used for grazing, this practice should continue to be an important use after reclamation. A grazing study was initiated in the spring of 1976 on reclaimed land to evaluate the response of reclaimed soils and vegetation to three grazing intensities: light, moderate and heavy.

Since grazing has been studied for only one season, no cumulative effects have occurred as yet. First year grazing did not cause any apparent cattle health problems, and gains were comparable with those obtained from unmined grazing studies with similar plant and cattle species and stocking rates (20). By the end of the first grazing season, the effects of the three grazing intensities were evident, in the amount of vegetation that remained, and subsequent grazing seasons should provide opportunities to observe the response of both soil and vegetation.

All of the current studies discussed herein have been designed to evaluate the time factor on reclamation. Therefore, these studies will be continued for 3 to 5 years, or longer. Changes in the soil, spoil, and vegetation over time will be monitored to provide a means for evaluating the effect of different weather patterns on the production of both annual crops and perennial plant communities over several years. Conclusions from past and current research projects could change as more conclusive data are collected and analyzed in the future.

REFERENCES

(1) Wali, Mohan K., (editor), *Practice and Problems of Land Reclamation in Western North America,* The University of North Dakota Press (1975), 196 pp.

(2) Brummet, R.W., and O'Donovan, J.D., in press, "Technology and Economics of Mining Texas Lignite," in Kaiser, W.R., ed., *Proc. 1976 Gulf Coast Lignite Conference,* Univ. Texas, Austin, Bur. Econ. Geology.

(3) Ellsworth, K., 1975, "Electrification: Two New Texas Coal Lines Are Coming in Under the Wire," *Railway Age,* vol 176, (No. 24), p. 22.

(4) Peretti, K.H., in press, "State of Technology of Deep Multi-Seam Mining, in European Opencast Mines," in Kaiser, W.R., ed., *Proc. 1976 Gulf Coast Lignite Conference,* University of Texas, Austin, Bur. Econ. Geology.

(5) Price, G.C., Manula, C.B., and Venkataramani, V., 1973, *Materials Handling Research: the Bucket-Wheel Excavator,* U.S. Bureau of Mines Information Circular 8580.

(6) Edgar, T.F., Kaiser, W.R., Thompson, T.W., Gray, K.E., and Humenick, M.J., 1977, *In Situ Conversion of Texas Lignite to Synthetic Fuels,* semiannual report (No. 4) to NSF-RANN, Univ. Texas, Austin, Dept. Chem. Eng., pp 10-49.

(7) Henry, C.D., Kaiser, W.R., and Groat, C.G., 1976, *Reclamation at Big Brown Steam Electric Station near Fairfield, Texas: Geologic and Hydrologic Setting,* Univ. Texas, Austin, Bur. Econ. Geology Res. Note 3.

(8) Hons, F.M., Askensay, P.E., Hossner, L.R., and Whiteley, E.L., in press, "Physical and Chemical Properties of Lignite Spoil Material as It Influences Successful Revegetation," in Kaiser, W.R., ed., *Proc. 1976 Gulf Coast Lignite Conference,* University of Texas, Austin, Bur. Econ. Geology.

(9) Lentz, R.C., 1974, *Relation of Eocene Depositional Environments to Sulfur Content and Quality of Surface Waters at Lignite Strip Mines near Fairfield, Texas,* Univ. Texas, Austin, M.A. thesis.

(10) White, R.L., 1975, "Land Reclamation in Texas—An Opportunity," in Wali, M.K., ed., *Practices and Problems of Land Reclamation in Western North America,* Grand Forks, ND, University of North Dakota Press, pp 154–169.

(11) Kaiser, W.R., 1974, *Texas Lignite: Near-Surface and Deep-Basin Lignite Resources,* University of Texas, Austin, Bur. Econ. Geology Rep. Inv. 79.

(12) Sandoval, F.M., Bond, J.J., Power, J.F., and Willis, W.O., 1973, "Lignite Mine Spoils in the Northern Great Plains—Characteristics and Potential for Reclamation," *Symposium on Research and Applied Tech. of Mined-Land Reclamation,* Pittsburgh, PA, pp 117–133.

(13) Power, J.F., Ries, R.E., Sandoval, F.M. and Willis, W.O., 1975, "Factors Restricting Revegetation of Strip-Mine Spoils," *Proc. Fort Union Coal Field Symposium,* vol 3, Clark, W.F., (ed.), Mont. Acad. of Sci., Billings MT, pp 336–346.

(14) Ries, R.E., Sandoval, F.M., Power, J.F. and Willis, W.O., 1976, "Perennial Forage Species Response to Sodium and Magnesium Sulfate in Mine Spoils," *In Proc. Fourth Symposium on Surface Mining and Reclamation,* National Coal Association, Washington, DC, pp 173–183.

(15) Power, J.F., Bond, J.J., Sandoval, F.M. and Willis, W.O., 1974b, "Nitrification in Paleocene Shales," *Science,* 183:1077–1079.

(16) Doering, E.J. and Willis, W.O., 1975, *Chemical Reclamation of Sodic Strip-Mined Spoils,* USDA-ARS-NC-20.

(17) Power, J.F. and Sandoval, F.M., 1976, "Effect of Sampling Method on Results of Chemical Analysis of Overburden Samples," *Mining Cong. J.,* 62:37–41.

(18) Sandoval, F.M. and Power, J.F., 1977, "Laboratory Methods Recommended for Chemical Analysis of Mined-Land Spoils and Overburden in Western United States," USDA Bulletin (in press).

(19) Barker, R.E., Ries, R.E. and Nyren, P.E., 1977, *Forage Species Establishment and Productivity on Mined Land,* ND Agr. Exp. Sta. Farm Res. (in press).

(20) Hofmann, L., Ries, R.E., Power, J.F. and Lorenz, R.J., 1977, "Grazing Study on Reclaimed Strip-Mined Land," *30th Ann. Meeting Soc. Range Mgt.,* p 18.

TRANSPORTATION, STORAGE, DRYING AND PULVERIZATION OF LIGNITE

The material in this chapter has been based on:

"Technology and Use of Low-Rank Coals in the U.S.A.," by G.H. Gronhovd and E.A. Sondreal of the Grand Forks Energy Research Center, Energy Research and Development Administration, prepared for the Seminar on Technologies for the Utilization of Low Calorie Fuels held in Varna, Bulgaria, April 20-22, 1976

Technology and Use of Lignite, Proceedings of the Bureau of Mines-University of North Dakota Symposium, Grand Forks, ND, April 27-28, 1967 (Bureau of Mines Information Circular IC 8376), compiled by W.R. Kube and J.L. Elder for the U.S. Department of the Interior Bureau of Mines, May 1978

Technology and Use of Lignite, Proceedings of a Symposium Sponsored by the U.S. Energy Research and Development Administration and the University of North Dakota, Grand Forks, ND, May 14-15, 1975 (NTIS GFERC/IC-75/2), compiled by W.R. Kube and G.H. Gronhovd, Cochairmen, for the U.S. Energy Research and Development Administration, 1975.

INTRODUCTION

Transportation

Historically the use of low-rank coals in the U.S.A. has been limited by the fact that the major deposits are located relatively far from the industrial centers. In recent years the demand for low-sulfur fuels has greatly changed the market patterns, and subbituminous coals from Montana and Wyoming are now being shipped to powerplants over distances as great as 1,500 miles (2,400 km). Because of their high moisture contents the lignites are still considered a regional fuel.

There are currently four alternative methods for supplying energy from low rank coal to distant load centers:

(1) Mine-mouth generation with electric power transmission—This method is quite common, especially when using the high moisture lignites. A disadvantage of this system is that many of the large coal deposits in the Powder River Basin (southern part of Northern Great Plains Coal Province) are located in areas where water is in short supply.

(2) Rail—This is presently the most common method of transporting U.S. low-rank coals, whether by individually scheduled railroad cars, with capacities commonly 60 to 120 short tons, or by unit trains, dedicated as multicar units exclusively to coal movement from a single source to a single discharge point. A typical situation would consist of a 10,000 ton unit train, having one hundred 100 ton cars, shuttling back and forth over a distance of 1,500 miles.

(3) Rail and barge—Where water transportation is feasible for part of the route, considerable savings can be realized by combining unit train and barge shipment. An example of this is a system now under construction to transport 8 million tons per year of subbituminous coal from the Decker Mine in Montana to the Saint Claire power generating plant near Detroit, Michigan. The system consists of 600 miles (965 km) of rail and 650 miles (1,045 km) of barge transport.

(4) Slurry pipeline—The first large scale coal slurry pipeline in the U.S.A. was the 108 mile long, 10 inch diameter system in Ohio, which was first operated in 1957. Although successful, operation has since been discontinued because of reduced railroad transportation rates in this area. More recently a 273 mile, 18 inch diameter, coal slurry pipeline went into operation in Arizona, transporting subbituminous coal from the Black Mesa Mine to the Mohave powerplant.

The coal slurry pipeline has favorable economics when very large tonnages are to be transported for long distances. It has environmental advantages in causing minimum disturbance to the surface areas, but the water requirements are a serious disadvantage in arid regions. No plans have been announced for the hydraulic transportation of lignite. Small-scale slurry transportation studies at GFERC indicated that lignite decrepitates more than bituminous coals and soluble humic acids are released, causing dewatering problems because of the tendency to form colloidal suspensions and gels. In addition, the low heating value of lignite would reduce the efficiency of transportation by slurry pipeline, because a 1:1 water to coal ratio is required for all coals.

No one method of transportation is likely to become preeminent in the use of low rank U.S. coals, and it is quite probable that expansion will be required in all methods in order to meet the predicted needs in the future.

Storage

Low rank coals with high inherent moisture content are generally more reactive than higher rank coals (1). Such reactivity is evidenced by a tendency to spontaneous heating during storage. However, experience has shown that the lignites and other low rank coals can be safely stored for long periods of time if proper precautions are taken to restrict air movement into the piles. One of the best examples of successful long-term large-scale storage is at the site of the Garrison Dam in North Dakota (2). Over two million tons of lignite were mined and placed in storage during construction of the dam.

The procedure was to build the stockpiles in 1 ft layers and then to compact with heavy rollers to about 70 lb/ft^3 (1,120 kg/m^3). Some of the piles are 40 feet high, and the edge slopes are about 14°. The stockpiles have remained in good condition for about 23 years with no serious problems. This procedure is now standard practice for construction of dead storage piles at powerplants.

Low rank coals decrepitate rapidly when exposed to air, and dusting can be a serious problem when handling or storing these coals in urban areas.

Drying

The high moisture content of lignite and other low rank coals can be detrimental in the following ways:

(1) It increases cost of transportation per heating value unit.

(2) It reduces efficiency of boilers.

(3) It causes derating when burned in boilers designed for higher rank coals.

(4) It causes freezing of coal in railroad cars in cold northern climates such as are experienced in North Dakota.

Because of these problems there is considerable interest in the study of low rank coal drying, especially now when the coals are being shipped great distances to meet sulfur regulations. At the present time, only one power generating installation is known in the United States where low rank coal (lignite) is dried before being burned. Installed capacity of the dryers at this plant is 450 tons/hr (3)(4).

Grand Forks Energy Research Center (GFERC), Grand Forks, N.D., is involved in a cooperative research program with industry to study the drying and handling, transportation, and storage of dried lignite and subbituminous coals (5). The test program has included the shipment of raw coal from mines in North Dakota and Montana to a commercial size dryer in Illinois, where the coal was dried before being shipped to the Research Center for stockpiling. The dryer was a drum dryer 11.5 feet (3.5 m) in diameter and 42 feet (12.8 m) long. Data on the drying tests are given in the table shown on the following page.

Oil spraying at the rate of 1.5 to 2.0 U.S. gallons/short ton (6.3 to 8.3 liters/metric ton) was found to be effective in reducing reactivity and controlling dust emissions while handling the dried coals. After drying, the coals were shipped 800 miles in open top cars to the GFERC. Only minor problems of spontaneous heating were experienced and those were in cars which had poorly sealed doors. The dried coals were placed in storage using layer buildup and compaction. The two 400 ton stockpiles have been carefully observed for 1½ years with no evidence of spontaneous heating.

	.... Coals Tested in Dryer	
	Subbituminous	Lignite
Initial moisture content, %	26	39
Initial higher heating value, Btu/lb (kcal/kg)	8,420 (4,680)	6,420 (3,565)
Final moisture content, %	16	22
Final higher heating value, Btu/lb (kcal/kg)	9,840 (5,465)	8,450 (4,695)
Initial coal size, inches	1.5 x 0	3 x 0
Inlet gas temperature, °F (°C)	810 (432)	825 (440)
Outlet gas temperature, °F (°C)	147 (64)	152 (67)
Coal temperature, inlet, °F (°C)	121 (49)	47 (8)
Coal temperature, discharge °F (°C)	145 (63)	146 (63)
Heat requirements		
10^3 Btu/short ton of water removed	380	540
10^3 kcal/metric ton of water removed	106	150

Moisture determinations on the lignite and subbituminous piles after 14 months of storage showed 26 and 14%, respectively, compared to 23 and 16% when placed in storage.

The results of these first-ever large scale drying, transportation, and storage tests on low rank U.S.A. coals have been quite encouraging, although no decision has been announced as yet to adopt commercial scale drying at other powerplants.

COMMERCIAL SCALE DRYING OF LIGNITE

Introduction

Large quantities of relatively low sulfur lignite and subbituminous coal occur in western United States. These low rank coals contain 20 to 40% moisture by weight and have a heating value of 7,000 to 8,500 Btu/lb. Removal of part or all of this surface and inherent moisture by drying increases heating value per pound, provides a product that is superior for powerplant utilization or gasification, and reduces shipping costs.

Drying these coals creates a product that is hot, dusty, and reactive to oxygen. Shipment and handling of dried coal is expected to be difficult because of the dust nuisance and the tendency toward spontaneous combustion during transit and storage. Methods to overcome these difficulties have been studied in the pilot plant at the Grand Forks Energy Research Center and it has been shown that these problems could be minimized. Commonwealth Edison Company and the Grand Forks Energy Research Center jointly conducted large scale experiments in which lignite was dried in a commercial scale dryer and shipped from Pekin, Illinois, to Grand Forks, North Dakota.

General Test Procedure

Dryer Test Facilities: The coal drying facilities are located at the Powerton Generating Station, Pekin, IL. A rotating drum type dryer was installed to remove surface moisture from washed coal during winter to aid plant operation and to serve as a facility for producing dried coal for testing purposes.

The rotating dryer drum is 11 feet 6 inches in diameter, 42 feet long, and the bed normally contains 14 tons of coal during operation. Coal residence time is 20 minutes at a production rate of approximately 40 tons/hr when removing 10% surface and inherent moisture. The auxiliary equipment includes a portable conveyor system to load coal into a steam jacketed feed hopper, a belt conveyor to move coal from the dryer into the normal plant coal handling system and an oil spray system. A vibrating screen may be used to separate a selected oversize fraction of feed coal which may subsequently be recombined with the product at a belt junction.

The hot drying gas, generated by combustion of fuel oil is tempered with air and forced through the dryer by an ID and a forced draft fan. Dust in dryer exhaust gases is collected by two sets of bag filters. This dust is sluiced to a waste pond.

For plant operation, the dried product is conveyed from the dryer to a belt transfer point and into the boiler house bunkers. For the tests described in this paragraph, the dried coals had to be moved on a series of three conveyor belts including two transfer points and an empty surge bin to the rail siding. An auger at the end of the last conveyor transferred the dried coal to a chute which discharges into a rail car. At each junction point, the coal dropped 5 to 10 feet between belts. The conveying system was 1,380 feet long and 2.7 minutes was required for transfer from dryer to rail car.

The system provided to spray the dried product with heated oil was installed above the belt conveyor at the dryer discharge. The type of oil used was Bunker "C", with a viscosity of 6,084 SUS at 100°F, a heating value of 147,563 Btu/gal, and is classified as No. 6. The oil was heated to 240°F and applied at pressures up to 110 psi.

Test Program: In the preliminary phase six rail cars of lignite were dried from 39% to 22% moisture content range and each was subjected to a different combination of treatments. These treatments included oil spraying, mixing of undried large particles with dried smaller fractions, and simulation of covered rail cars by topping with plastic sheeting. Cooling, in itself a treatment, occurred during conveyance and was constant for all tests. One car, as a standard, was loaded with untreated dried coal. The cars were observed several weeks in the station yard. Observations from preliminary tests showed that the most practical treatment was oil spraying combined with the cooling which occurred.

Using the procedure shown in Figure 4.1, 400 tons of dried lignite was produced, treated, and shipped from Powerton Station to GFERC for long term storage observations.

Lignite Tested: The test lot of lignite was obtained from the Gascoyne mine, Bowman County, North Dakota. The moisture content of Gascoyne lignite is typically higher than the average for lignites and has a comparatively "soft" structure. The proximate and ultimate analyses of lignite as charged to the dryer are shown in Table 4.1.

Figure 4.1: Schematic Process Used for Drying and Treating Coal

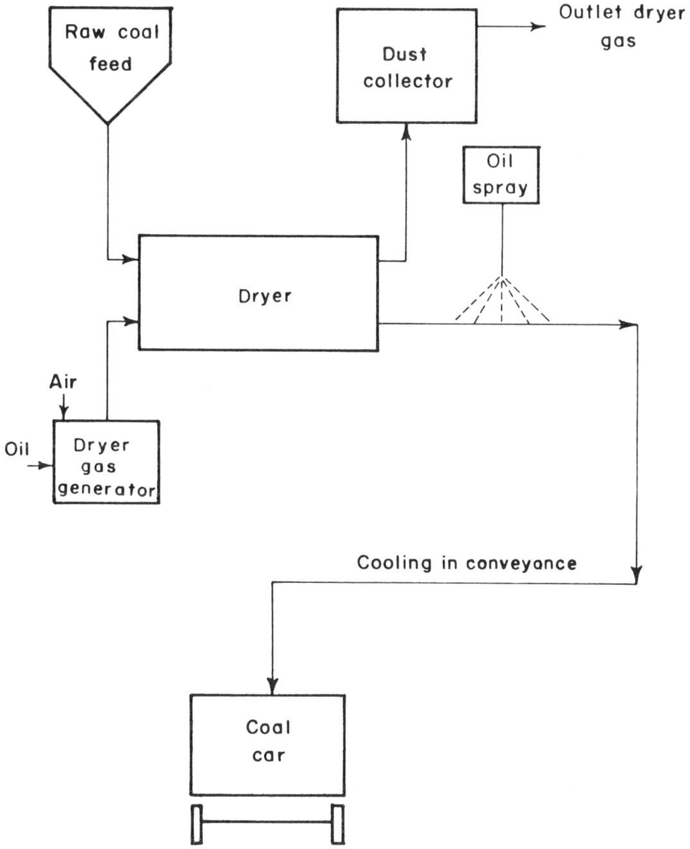

Source: NTIS GFERC/IC-75/2

Sampling Procedure: At the dryer, 20 pound sample increments were collected every 45 minutes from both feed and product conveyors. Dust collected from exhaust gases was sampled by collecting a section of the dust stream discharging from the bag filters. At the car, samples were periodically collected at the discharge chute.

Coal samples were placed in double layer plastic bags and returned to GFERC for analysis. The moisture content of each individual sample was determined but size consistency, and proximate and ultimate analyses were determined on composite samples. Special product samples were collected to determine the moisture content as a function of size.

Table 4.1: Analyses of Lignite Tested, As-Received Basis

	Gascoyne Lignite Coal
Proximate analysis, %	
Moisture	39.0
Volatile matter	25.2
Fixed carbon	25.9
Ash	10.0
Total	100.0
Ultimate analysis, %	
Hydrogen	6.8
Carbon	38.7
Nitrogen	0.5
Oxygen	43.0
Sulfur	1.0
Ash	10.0
Total	100.0
Heating value, Btu/lb	6,420

Source: NTIS GFERC/IC-75/2

During unloading of each car at GFERC, an increment of the coal discharging from the under-car conveyor was collected every 5 minutes, and the resultant sample processed for analysis.

Test Drying of Lignite

The dryer facility had not been operated for an extended period of time prior to the tests. The drying gas flow rate was found to be limited to about 65% of design rate due to a problem involving the flame stability of the fuel oil burner.

Test lignite was shipped by rail from the mine in North Dakota to Powerton, unloaded at the plant storage pile by the normal station procedure, and transferred to the dryer site by truck. At the dryer site, a front end loader dressed the lignite into a cone shaped pile and also hauled the lignite from the pile to the feed hopper during drying operations.

For the shipment tests the lignite was dried from 39.0% to 22.0% moisture at a rate of 15.4 tons/hr. The low product rate for lignite was caused by high moisture and large particle size in the feed. For lignite, heat requirement for drying was 5.4 MM Btu/ton of water removed. The feed size of test lignite was 1½ by 0 inches and 3 by 0 inches, as is normally produced at the mine. A total of about 400 tons of dried product was produced in the transportation test phase.

Particle Size Degradation: The removal of moisture from a lignite produces significant changes in its physical characteristics. As moisture is removed, the particle shrinks producing cracks and fissures. This process, called "weathering" or "slacking" creates a friable material. The drying and handling therefore reduces size of particles and creates dust. The extent of this process is related to the degree of drying, the original size of the particle, and the petrographic structure of the lignite.

Moisture Content as a Function of Particle Size: The relationship between coal particle size in product and the degree of drying was determined. Immediately after drying a sample of product was separated into respective size fractions. Each fraction was placed into individual plastic bags and analyzed for moisture content by the ASTM method. The variation in moisture content as a function of size for lignite is shown in Figure 4.2.

Figure 4.2: Moisture Content of Dried Lignite Coal Product as a Function of Size

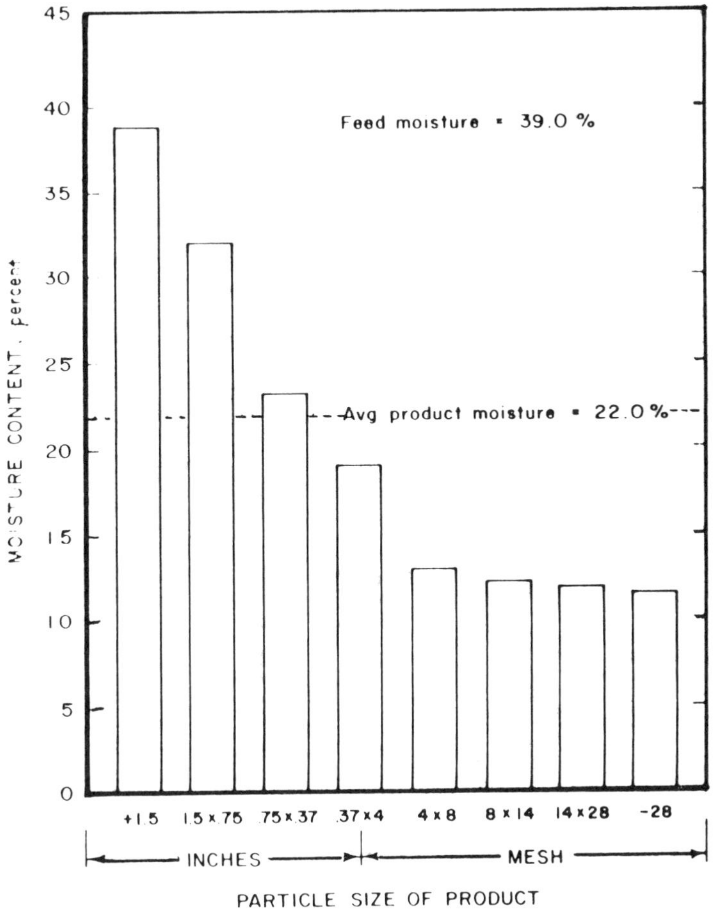

Source: NTIS GFERC/IC-75/2

In lignite, as shown in Figure 4.2, the +1½ inch size in the product had not lost significant moisture, but the 28 mesh by 0 fraction was dried to 12.0%. All 3/8 by 0 inch particles had moisture contents less than the average while larger particles were higher. Data show that larger particles in the product were not dried to any great extent and perhaps existed for that reason.

Production of Dust in Drying: The particle degradation associated with drying, and subsequent handling produces dust. The dust produced in the dryer was removed from the exhaust gases by two sets of bag filters in series, and was sluiced to a waste area near the test facility. In a commercial operation, it would undoubtedly be used as fuel to heat the drying gases. The amount of dust generated in drying was estimated by weighing the material discharged from the collector for several specific periods and projecting these data to an hourly basis. The particle size and moisture content was also determined.

The amount of dust produced during drying the Gascoyne lignite amounted to 7.7% by weight of the feed on a moisture free basis. Moisture content was 7.5% and approximately 40% was less than 200 mesh size. "Uncompacted" and "compacted" densities for lignite dust were 40.4 and 50.6 lb/cu ft, respectively.

Handling and Shipment Characteristics

Oil Spraying: The preliminary tests demonstrated that control of dust nuisance was required regardless of any other consideration.

In the preliminary tests, oil was applied to the dried lignite, but due to improper spray nozzles the oil was squirted on the coal rather than sprayed on it. In addition, the rate of application could not be properly controlled. As a result, coverage was inadequate, and the oil loss excessive. About 10 gallons/ton of product was used.

For the transportation phase, the oil spraying system was modified to improve particle coverage and application. Oil spray rate was decreased for each successive car of dried coal to determine the minimum quantity required. Also one car was loaded with undried coal from the stockpile which was not oil treated.

Seven cars were loaded with dried lignite, one of which was also not oil treated. The oil application rate for the cars of treated lignite ranged from 1.0 to 1.9 gallons/ton of product. The effectiveness of application was again improved compared to previous testing by installation of different nozzles; however, even so the spraying procedure was not optimized. Despite these inadequacies, the dust nuisance at the cars was significantly decreased.

Density of Lignite in Car: After loading, each car was leveled with hand rakes. The average density of dried lignite at Powerton and Grand Forks was 47.0 and 51.1 lb/cu ft. These values are comparable to bulk densities of prepared as-mined lignite coals.

Oxygen Content of Pile Gases in Car: The oxygen content in the void spaces between particles in the loaded cars was measured with a portable oxygen analyzer. A pipe probe was inserted into the lignite in the car immediately after the car was loaded. Gas samples were periodically withdrawn and oxygen content determined. The oxygen content of the void gas in dried lignite decreased from 21 to 1% within one hour.

Results show that the oxygen is depleted from pile gases very rapidly. Therefore, any additional oxygen for reaction must be introduced into the lignite from outside the car. The rate of oxygen depletion in the cars is also indicative of the high reactivity of the dried lignite as loaded even with oil treatment and cooling.

Measurement of Dust Loss During Transit: An effort was made to compare the amount of lignite lost by particle entrainment and wind blown dust from open car tops during shipment of treated and untreated lignite. For this purpose, special dust collectors were installed on test cars.

Each dust trap consisted of a box with openings at both ends through which air could flow. Inside, several baffles were installed to deflect the air flow containing entrained dust from the front opening through and over an oil bath, and out the rear opening. The exit opening was directed downward so that air and dust from other cars could not enter. One collector was installed at each end of a test car. For either direction of travel, a sample of the air passing over the top of the car entered the collectors. Dust and entrained particles were removed and collected. One of the traps together with a rain gauge was installed to measure precipitation during transit. At the GFERC the collectors were removed, the oil containing the particles recovered, diluted with solvent, and filtered.

For the lignite tests, collectors were installed on a car of as-mined coal from Gascoyne, ND to the Powerton Station, Pekin, IL and on one car of untreated and treated dried coal transported from Powerton to GFERC. Dust loss determined in shipment from Powerton to Grand Forks cannot be compared directly to the data for as-mined coal in shipment from Gascoyne to Powerton but does illustrate that significant losses do occur.

For dried lignite, the quantity of dust collected on the untreated car was six times that of the oil-treated. The quantity of dust collected during transportation of the raw coal was less but still twice that collected from the dried oil-treated coal despite the fact that the size of as-mined coal was 3 by 0 inches while the size of the dried treated coal was much smaller. The largest particle size of the material in the dust collector was retained on a ¼ inch screen.

Effect of Air Leakage in Cars: Heating of lignite in rail cars or in stockpile can occur when oxygen supply for reaction with the lignite is provided. If oxygen is prevented from entering the car or pile, the reaction is limited to that which occurs between the coal and the initial pile gases. For a rail car, in particular, openings by which air can enter at the car bottom and create a flow through the lignite bed, is potentially a major source of heating problems.

The extent of ignited lignite, proved to be very localized, and limited to a depth of several inches in the vicinity of the door closure. Ignition occurred also in the car of undried lignite as well as the cars of dried lignite.

As a result of this experience extensive precautions were taken to prevent air leakage during the transport of dried lignite. Openings at hopper doors were filled with fiber insulation and covered with plastic sheeting before loading. Ignition at door openings did not occur during this shipment.

Unloading and Stockpiling: At Grand Forks, the dried lignite was unloaded using an under-car conveyor and hauled to the stockpile site with trucks. Each load was successively spread on the pile in layers to cool. Compaction and operation on the pile could not be accomplished with a variety of small equipment but a large front end loader successfully operated on the dried lignite to spread and compact the pile.

In contrast to the problems of ignition which occurred during the shipment, an unloading problem was encountered because of freezing of coal to car walls. As loaded, the temperature of the dried lignite exceeded ambient temperature, and thus was releasing moisture as water vapor. After loading, the coal continued to release steam which condensed on cold car walls and adjacent particles. As a result, a layer of particles, agglomerated by freezing, were formed at the car wall. Personnel had to enter the car and manually dislodge the coal during unloading. Agglomeration by freezing provides additional justification for cooling of the coal before loading.

Summary of Tests and Conclusions

The test lot, exceeding 400 tons of lignite was produced in a commercial scale dryer and shipped in standard bottom dump open top cars a distance of 750 to 800 miles and stockpiled. The lignite was upgraded from 6,420 to 8,450 Btu/lb.

Physical characteristics of the particles are altered in drying low rank coals resulting in size degradation and generation of dust. Extent of drying is affected by feed size and the larger the particle, the less drying occurs at given operating conditions. The smaller the particle size, the more effective is the drying operation. The amount of dust produced and collected from exhaust gases is appreciable and if utilized to heat dryer gases will provide all or a substantial portion of heat required for drying.

The dried product is hot but appears to cool rapidly by open conveying or exposure. With lower product temperatures, the coal has less tendency to heat spontaneously in subsequent shipping or storage. In winter shipment particularly, the temperature of dried coal must be reduced before storage or transport to prevent agglomeration by freezing because moisture as a vapor is released from particles until it is cooled.

Regardless of any other considerations, dried low rank coals must be treated to reduce dust nuisance in subsequent handling operations unless a completely closed system is provided. Application of 1.5 to 2.0 gallons/ton of heavy oil was found to be of significant benefit in respect to dust control and reduction in wind losses during transport.

The bulk density of low ranked coals is not altered by the drying process. Because the Btu/lb value is upgraded with drying, a specific volume of dried coal will have a higher heating value than the same volume of raw coal. The oxygen of the pile gases within a car is rapidly consumed by the dried coal to provide an inert atmosphere. If no additional air is provided, additional temperature increases cannot occur. The necessity for eliminating opportunity for an oxygen supply at car bottoms was demonstrated. Air penetration into the top surface of coal appears to be insufficient to generate sufficient heat for ignition.

Stockpiling and handling requirements do not appear to be significantly different from as-mined coals. The need for careful and thorough compaction is somewhat greater but normal coal handling equipment can be utilized.

LIGNITE PULVERIZATION

Introduction

This section will survey the status of lignite pulverization for use in pulverized fuel fired electric generating plants in the Northern Plains area of the United States. It will describe the investigations conducted at the Grand Forks Energy Research Center on pulverizing lignite.

The first installation of a boiler fired by pulverized lignite in its consuming area was in 1948, at the Crookston, MN, station of the Otter Tail Power Co. (6). By present standards of size, it would be a very modest beginning. Of the two boiler units installed in the station, one was provided with pulverizers. This unit was the forerunner for the larger centralized generation units needed by the expanding power systems. It demonstrated to the lignite industry that (a) lignite, as-mined, could be successfully pulverized; (b) it could be readily ignited at a degree of fineness comparable with or less than that required for higher rank coals; and (c) existing pulverizing equipment used for higher rank coals could be used to pulverize lignite. The principal requirement was to sweep the mills with preheated air to partly dry the lignite as it was pulverized.

Status of Pulverized Firing Method

From the standpoint of capacity and generation, the pulverized method of firing currently dominates the lignite consuming industry. There is no doubt that lignite can be successfully pulverized, but some doubt has been expressed that the technology of lignite pulverization has advanced sufficiently to cope with future conditions.

In the first installed units, design was conservative and the size and number of pulverizers provided what was considered to be an ample reserve. During this period, the load requirements were such that the capabilities of the units were reached only for brief periods each day and load factors were not high. Problems relating to capacity of pulverizers and size of drive motors were seldom encountered. The natural tendency was then to design more closely to the expected needs and reduce reserve capacity to a minimum. Gradually, however, with the growth of power requirements, the loads of the larger pulverized fired stations have increased, and the capabilities of the units are maintained for longer periods.

With this progressive increase in electrical loading, new areas in the producing mines were opened and difficulties have been encountered. Lignites that are more difficult to grind have been mined and used in equipment designed to operate with the types of lignite encountered in previous years. Pulverization of lignite is not as simple as was originally supposed.

A number of examples will illustrate these difficulties. One unit has been found to operate very successfully on lignite from the major seam of the mine which supplies its fuel. When lignite from a seam directly beneath the major seam is received, the pulverizers cannot deliver enough fuel to maintain boiler capacity despite almost identical heating values for the two lignites; in another case, the use of lignite that had been frozen in storage or transit was found to limit the capability of the boiler unit because the pulverizers could not deliver enough fuel; in a third case, the pulverizers installed in a plant could not provide sufficient fuel to maintain capacity loading apparently because the indicated grindability of the lignite was greater than the true grindability.

These problems have been tolerated or solved, but pulverization of lignite is now viewed with more respect and attention. The need to restore the initial confidence in pulverized firing has never been greater and the progress made will depend upon the cooperation of all segments of the industry.

At this time, pulverizers manufactured by three different companies have been used on lignites in commercial units. Grinding actions are employed which exert compressive force on individual particles. The manner in which this force is applied differs: one utilizes rollers which are mounted inside a rotating bowl and particles are pulverized between the rollers and the inside edge of the bowl; a second one has a rotating race with a number of balls which ride on and over particles to be pulverized; and the third utilizes a rotating table on which massive wheels rotate to ride on and over particles to be pulverized. A fourth type (not used at present) is an attrition and impact type in which particles were bounced and crushed in suspension between a series of pegs or teeth.

All operating mills pulverize natural as-received lignite as feed. They are provided with internal classification for return of oversize for additional grinding and utilize preheated sweeping gas to transport and dry the lignite as it is ground, classified, and transported to the burners. The lignite as burned contains 25 to 30% moisture, and the fineness is in the range of 55 to 65% through a 200 mesh sieve. It has been found that lignite need not be as fine as bituminous coals owing to the superior ignition characteristics of lignite. In general, wear or maintenance requirements have not been excessive, and are less than was expected.

There are differences between pulverizers manufactured by the various companies. It is not the purpose of this discussion to compare one pulverizer with another; the important thing is that each of the major boiler manufacturers has operated its specific equipment on lignite. In addition to the units currently operating, tests were recently conducted in a boiler unit at Alma, WI, in which lignite was successfully pulverized in a ball mill that normally operates on bituminous coals.

Lignite can be pulverized in a variety of pulverizers, but it must be recognized that the technology has not advanced to the point where the industry has solved all of the problems that can develop in current practice nor have concentrated efforts been made to pulverize lignite at optimum conditions.

Research in Pulverization of Lignite

At the Grand Forks Energy Research Center, research projects in pulverization of lignite have been conducted on a continuing basis for years. These have included the study of lignite grindability as determined in the Hardgrove grindability machine, tests and observations of commercial units, and study of different types of pulverizers in the pilot plant (7)(8)(9)(10).

Grindability of Lignites: The standard laboratory method used by the power industry to compare the grindability of coals is the Hardgrove Grindability Test Method. It was developed for and used with success to predict pulverizer operation on bituminous coals. The first investigation into pulverization was to determine the applicability of this test method to lignites and establish, if possible, the variation in grindability for lignites obtained from different mine sources.

Basically the test consists of placing a prepared 50 g sample of coal in a bowl in which it is subjected to the grinding action of steel balls rotated for a specified number of revolutions in a race in the bottom of the bowl. A fixed weight mounted on the upper race that rests on the balls provides a constant grinding pressure. The sample is then removed and sieved. The weight of sample which passes the 200 mesh sieve is determined, and an index is computed using a formula that relates the grindability of the sample to the grindability of a standard coal.

Based on accumulated experience and data, each manufacturer has developed performance curves by which he can relate the indicated Hardgrove index to the capabilities of specific pulverizers for a variety of conditions. Ultimately, a mere 50 g sample of the coal proposed for use in a powerplant is influential in establishing the size and number of the pulverizers that are installed in a powerplant which requires the investment of millions of dollars.

The study showed that the Hardgrove index determined on lignites was a reproducible value providing that the procedure and test conditions were identical for each step in the test. A very important factor in the determination of the grindability of lignites was the moisture content of the sample as it was tested. The moisture content of lignite is variable depending upon the particle size, humidity, and temperature to which it is exposed in the laboratory.

For the lignites tested in the study, the variation of the indicated Hardgrove index as a function of moisture content of the sample was a curve; high index at the natural moisture content decreasing to a minimum as moisture content was decreased to the midmoisture content of 15 to 25%, and increasing again as moisture was reduced to still lower values. Variation in grindability of lignites obtained from a number of mine sources was appreciable. The highest and lowest Hardgrove grindability indexes of the lignites studied are shown as a function of moisture content in Figure 4.3.

From these presentations, one would expect that operation of the commercial units on natural lignite would be at optimum conditions for pulverization. Based on all test results, it has been concluded that the indicated grindability at the higher moisture contents is very misleading and does not properly reflect the true grindability.

Little success has been obtained in correlating the Hardgrove grindability index to operation of pilot plant pulverizers. There are a number of reasons for this. Basically, the test is greatly influenced by the relative ease with which particles can slide upon each other moving into or away from the grinding zone in the mill. For lignites, the surface characteristics are greatly changed with moisture content. At higher moisture contents the particles tend to adhere to each other and to the grinding elements more readily than at lower moisture contents.

Figure 4.3: Hardgrove Grindability Index Versus Moisture Content of Two North Dakota Lignites

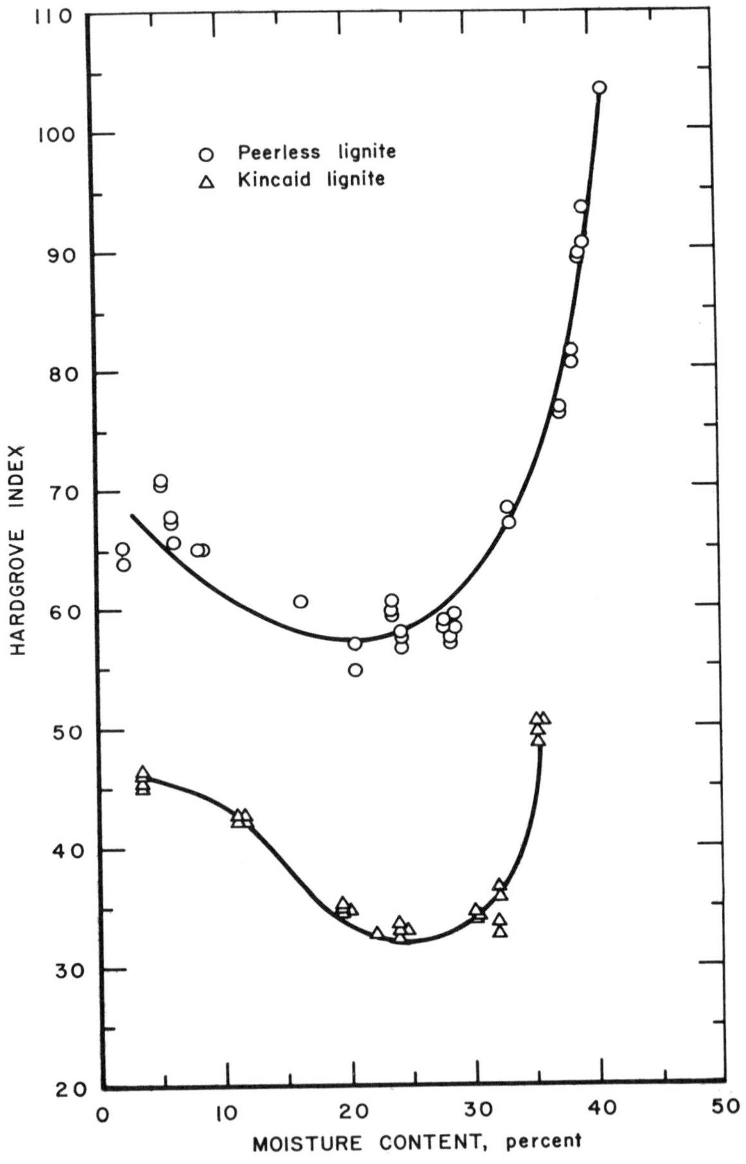

Source: IC 8376

In extreme cases the particles are carried along with the balls and some grinding occurs at the upper race. The surface characteristics of lignite are such that the grindability indexes determined at lower moisture contents are probably a more nearly true comparison for lignites. The unit also has no provision for removing the ground material during the test as occurs in airswept commercial mills; consequently, the index does not reflect the same grinding conditions as in the commercial mills.

The Hardgrove machine has been modified in an effort to obtain more information concerning the grinding properties of lignite. These modifications have included attaching a torque arm to the bowl to indicate the amount of energy expended for grinding, and measurement of the thickness of material in the grinding zone. In general, the energy required by the Hardgrove machine in a test corresponds to the amount of sample pulverized. The higher the indicated index, the higher is the total energy input. The thickness of lignite being ground (the "gap") does not correspond directly to the index as the gap is very high for those lignites that have low grindability index. The objective of this continuing study is the possibility that these modifications will permit the standard laboratory method to be of greater use to industry in application to pulverization of lignite.

Currently, the index is being determined at moisture contents of 35, 25, and 15% for samples of lignite collected at each active mine. At specific intervals on the exposed lignite bed, numerous auger drill samples have been collected. When this survey is completed, the variation in grindability of lignites as indicated by the Hardgrove index within producing mines and within specific producing areas will be established. Despite the apparent lack of confidence in the Hardgrove index for lignites, relative information is provided and perhaps with more experience and test work, more meaningful use of the index for lignites might be accomplished.

Grindability of Petrographic Constituents in Lignite: Lignite is a mixture of the remains of trees and plants, and its physical characteristics are related to those of the specific plant constituents. Petrographic data is limited; however in conjunction with the detailed petrographic study of one channel sample of lignite, the variation in petrographic composition was correlated to the variation in Hardgrove grindability index for the depth intervals of the seam samples.

The various petrographic components were separated from a sample of lignite and the grindability of each was determined. The fibrous component, anthraxylon, was found to have a grindability of 32. The translucent attritus was slightly higher in grindability with an index of 39, but the opaque attritus had a grindability index of 92. The grindability of the fourth petrographic component, fusain, was not determined because of its extreme softness and because it is present in lignites in negligible quantities. Based on the great difference in grindability of these components, the grindability of a sample of lignite could show a wide variation depending upon the relative proportion of these constituents.

The Hardgrove index is plotted versus vertical location in the seam for this sample of lignite in Figure 4.4. The total percent of the two components of low grindability, anthraxylon plus translucent attritus, is directly related to the Hardgrove index. The grindability of the seam shows a wide range with depth equivalent to or greater than the entire range found in samples of produced lignite.

Figure 4.4: Hardgrove Index and Percent Translucent Material Versus Depth in Seam

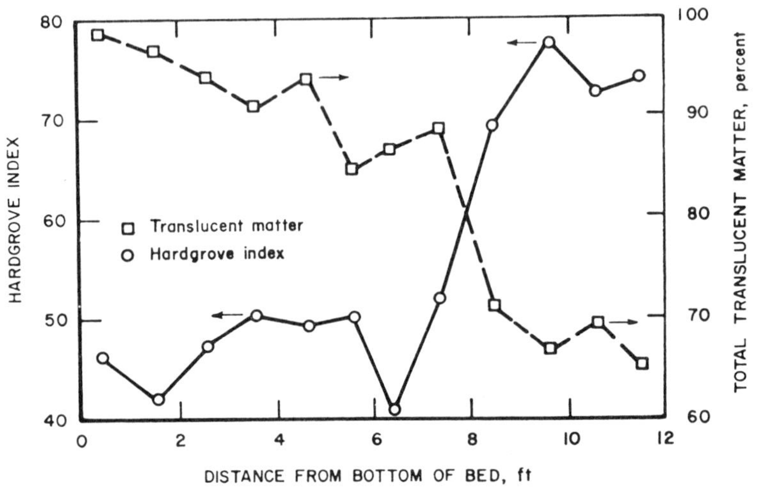

Source: IC 8376

On the basis of these data there is little reason to expect that the grindability of all lignites can be expressed by a single number. Design of a pulverized fired unit is made more difficult owing to the variable nature of the lignite grindability; therefore, a great number of samples must be collected and tested to establish the range of the characteristics of the lignite from any particular area.

Pilot Plant Tests of Pulverizers: Three pilot plant size pulverizers have been studied. These include a hammer mill (impact) pulverizer; a ball mill; and a ring-roller mill. The ring-roller mill is much smaller, but identical in construction to the pulverizers installed in the Stanton, ND, generating unit of United Power Associates.

Objectives of these studies have been to determine the effect of the variables associated with lignite pulverization on the power consumption and capacity of test mills. Variables include the source of the lignite and the moisture content of the lignite as pulverized. Two methods of reducing the moisture content of the lignite were studied: in-the-mill drying by the use of preheated sweeping gas, and predrying the lignite before pulverization.

The major variable associated with lignite pulverization was found to be the moisture content. Contrary to the indication of the Hardgrove index, power consumption is highest and capacity is lowest at natural moisture content. Power consumption decreased and the capacity of each mill increased as the moisture content of the lignite was reduced. This is true whether the lignite is predried or dried in the mill with preheated sweeping gas. In most cases the fineness of the product is also increased as the degree of drying is increased.

To provide a method of comparing individual tests, a "comparison factor" has been devised. This factor relates the fineness of product to the power consumption per ton of lignite pulverized. It can be expressed in several ways; the percent of minus 200 mesh product divided by the kilowatt-hours required per ton, or as pounds of minus 200 mesh product produced per kilowatt-hours required. A plot of this factor as pounds of minus 200 mesh product produced per kilowatt-hour versus moisture content of the tested lignite is given in Figure 4.5.

Figure 4.5: Pounds of Pulverized Lignite Produced per Kilowatt-Hour Versus Moisture Content of Lignites*

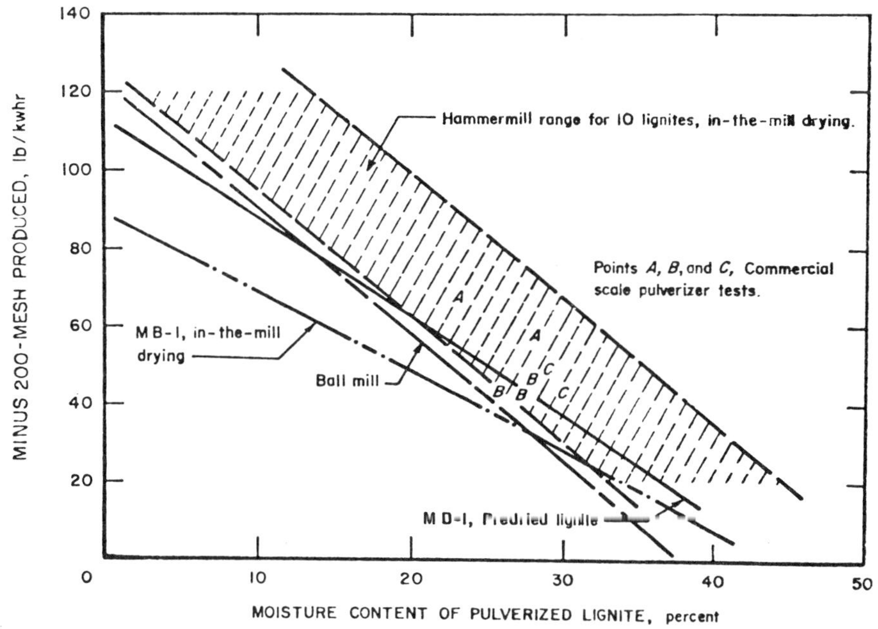

*Pilot plant and commercial scale test results

Source: IC 8376

Data collected and compiled from all pilot plant tests were used for this relationship, and the limited data collected at commercial stations are included for comparison. Three types of pilot plant mills and three different commercial mills are represented. The band representing the test results obtained in the hammer mill includes the entire range of variation for lignites from 10 mine sources. This range was also established by the same two lignites which established the range in Hardgrove grindability indexes shown in Figure 4.3. Test results obtained in the ball mill for both types of drying correspond very closely to the results obtained on the same lignite in the hammer mill. Results with the MB-1 mill for the tests on predried lignite also correspond very closely to results with the other two mills. For in-the-mill drying, results on the MB-1 indicate some sacrifice in performance. Results of these tests cannot be compared directly to the other two mills because procedures, classification, and residence times for in-the-mill drying could not be duplicated for all test mills. In particular, the internal classification system of this mill reduced the effectiveness of the in-the-mill drying compared to the external classification system used in the other two test mills.

The data are shown to illustrate that performance of a mill is directly related to the effectiveness of the drying that takes place within the mill.

Data which was obtained during operation on commercial sized units is also plotted. As could be expected, the product moisture contents range mostly from 25 to 30%, but it is significant that all of these points fall within the boundaries of the data obtained with pilot plant scale units.

The relationship between drying of lignite and the capacity of test pulverizers is shown in Figure 4.6.

Figure 4.6: Effect of Drying on Capacity of Pilot Plant Pulverizers

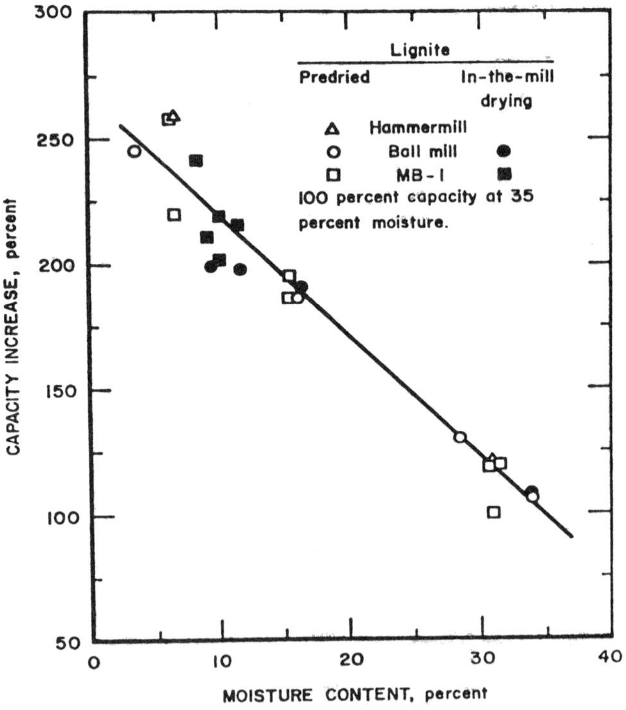

Source: IC 8376

Data have been plotted for each of the three mills in which the maximum output in pounds of pulverized product was established for a lignite as a function of its moisture content. If the lignite is dried to 15% moisture, each of the pulverizers could pulverize up to 75% more lignite without overloading. Drying to even lower moisture contents showed still further increases in capacity. It must be recognized that these increases are in weight units; the increase in capacity would be even higher if the heating value of each pound of pulverized product were considered.

LIGNITE PELLET-FORMING PROCESS

The Department of Energy, working with Babcock Contractors, Inc., has successfully demonstrated a new process for drying and compressing lignite coal into more economical and easily transportable pellets.

The process, called "Lignipel" (for "lignite pelletizing"), could significantly increase the use of low ash, low sulfur lignite in power plants farther from the mine by making such use more attractive economically.

Although lignite is one of the most abundant fossil fuels in the U.S., its use to date has been severely limited by its high moisture content–typically 35 to 40% by weight–which makes transportation costly.

Today, practically all lignite is consumed in power plants near the mines. Moreover, conventional drying at the mine creates a product that is dusty, prone to spontaneous combustion and to picking up moisture when left in the air.

The Lignipel process reduces or eliminates these problems by reducing the moisture content to about 15%–a level which results in the strongest material–and creating a hard pellet that is resistant to spontaneous combustion. The ¾ inch diameter pellets have a heating value of about 9,500 Btu/lb.

The process was developed by engineers at DOE's Coal Preparation Section at Bruceton, PA. Along with a team from Babcock Contractors, they have demonstrated the process at Allis Chalmers Process Research and Test Center in Oak Creek, WI. More than 50 tons of lignite pellets were produced during the demonstration, at rates of almost 5 tons per day.

The process basically involves an adaptation of existing coal preparation equipment. Crushed lignite is first fed into a mixer where an asphalt chemical is added to bind the particles for added strength and easier handling. The lignite then moves to a conventional disc pelletizer and finally to a specially adapted dryer.

As the pellets move through the dryer by conveyor belt, they are first dried by an updraft of heated air, then by a downdraft. By drying the lignite alternatively in two directions, more material can be dried in less time and the size of the dryer reduced, resulting in overall cost savings.

The process adds roughly $4 to $7 in costs per ton of product, which is comparable to conventional bituminous coal preparation.

Researchers will use the lignite pellets from the demonstration to test storage, transportation and burning characteristics. Also, as part of its $300,000 contract with DOE, Babcock Contractors is designing a 4,000 ton per day lignite pellet plant.

The Lignipel process could be in commercial use beginning in the early 1980s.

REFERENCES

(1) Sondreal, E.A., and Ellman, R.C., *Laboratory Determination of Factors Affecting Storage of North Dakota Lignite,* BuMines RI 7887, 83 pp (1974).

(2) Ellman, R.C., Belter, J.W., and Dockter, L., *Long-Term Storage of Lignite at Garrison Dam,* Riverdale, ND, BuMines RI 7037, 56 pp (1967).

(3) Parry, V.F., *Low-Temperature Carbonization of Coal and Lignite for Industrial Uses,* BuMines RI 5123, 27 pp (1955).

(4) Caracristi, V.Z., and Mumper, H.D., Combustion of Crushed, Dried Texas Lignite and Char in Steam-Power Boilers, presented at June 1959 meeting of Fuels Division, ASME, 15 pp.

(5) Ellman, R.C., Paulson, L.E., and Cooley, S.A., "Commercial-Scale Drying of Low-Rank Western Coals, Part 1—Rail Shipment Test Observations," presented at 1975 Lignite Symposium, Grand Forks, ND, 21 pp (May 14-15, 1975).

(6) "Pulverized Lignite Burned Successfully in Minnesota Powerplant," *Electric World,* Vol. 135, No. 11, pp 94-95 (March 12, 1951).

(7) Ellman, R.C., and Belter, J.W., *Grindability Testing of Lignites,* BuMines Rept. of Inv. 5167, 38 pp (1955).

(8) Ellman, R.C., and Belter, J.W., *Pulverizing Lignite in a North Dakota Powerplant,* BuMines Inf. Circ. 8032, 32 pp (1961).

(9) Ellman, R.C., Belter, J.W., and Dockter, L., "Pulverizing Lignite in a Pilot Plant Test Unit," *Proc. Am. Power Conf.,* Vol. 25, pp. 513-524 (1963).

(10) Dockter, L., Belter, J.W., and Ellman, R.C., *Pulverizing Lignite in a Ball Mill,* BuMines Rept. of Inv. 7010, 13 pp (1967).

LIGNITE COMBUSTION

The material in this chapter has been based on:

"Technology and Use of Low-Rank Coals in the U.S.A.," by G.H. Gronhovd and E.A. Sondreal of the Grand Forks Energy Research Center, Energy Research and Development Administration, prepared for the Seminar on Technologies for the Utilization of Low Calorie Fuels held in Varna, Bulgaria, April 20-22, 1976

Technology and Use of Lignite, Proceedings of the Bureau of Mines-University of North Dakota Symposium, Grand Forks, ND, May 9-10, 1973 (Bureau of Mines Information Circular IC 8650), compiled by G.H. Gronhovd and W.R. Kube of the Grand Forks Energy Research Center for the U.S. Department of the Interior Bureau of Mines, 1974

Technology and Use of Lignite, Proceedings of a Symposium Sponsored by the U.S. Energy Research and Development Administration and the University of North Dakota, Grand Forks, ND, May 18-19, 1977 (NTIS GFERC/IC-77/1), compiled by G.H. Gronhovd and W.R. Kube, Cochairmen, for the U.S. Department of Energy, 1977

Research and Development Relating to Lignite Use in Power Production (NTIS CONF-7606131-1), by E.A. Sondreal, G.H. Gronhovd and W.R. Kube of Grand Forks Energy Research Center, prepared for the Gulf Coast Lignite: Geology, Utilization, and Environmental Aspects symposium held in Austin, TX, June 1976

Fluidized-Bed Combustion Test of Low Quality Fuels (NTIS MERC/RI-78/3), by J.S. Mei, U. Grimm and J.S. Halow of Morgantown Energy Research Center, prepared for the U.S. Department of Energy, April 1978.

OVERVIEW OF LIGNITE COMBUSTION FOR POWER PRODUCTION

Methods of Firing

Most of the low-rank U.S. coal produced is burned in electric generating stations although some is burned in other industrial boiler applications. The most common firing method for central station power boilers burning low-rank coals is pulverized coal (pc); however, in recent years, three cyclone-fired boilers have been put into operation burning North Dakota lignite. For smaller industrial boilers, the usual firing method is spreader stoker with traveling grate ash discharge.

There is also much interest in development of fluidized-bed combustion for application to lignite firing (1). The problem of ash fouling and the problem of controlling sulfur emissions are both potentially solved by fluidized-bed combustion (FBC). FBC involves the combustion of minus ¼-inch coal at a controlled temperature of 1500° to 1800°F in a bed of inert material held in suspension (fluidized) by a flow of air up through the bottom of the bed. The relatively low temperature of the bed, lower than the fusion of the ash, should greatly reduce or eliminate the ash fouling problem.

Also, at this temperature, the sulfur dioxide generated from the coal can be absorbed on limestone, or in the case of lignite, at least partially on the alkaline coal ash. The capability of absorbing sulfur dioxide during combustion eliminates the need for flue gas desulfurization, which is the primary incentive hastening the development of FBC. Other envisioned advantages are a reduced emission of NO_x and the possibility of building smaller and less costly boilers because of the inherently high heat transfer rates between the bed and immersed heat transfer surfaces.

The DOE program on FBC has the objective of rapid development for both utility and industrial applications, with commercialization targeted for the 1980s. Atmospheric fluidized bed combustion (AFBC), which is farthest advanced, is a natural successor to pulverized coal firing for steam generation. Pressurized fluidized bed combustion (PFBC), at a less advanced stage of development, is envisioned to be applied to combined-cycle power generation provided that the hot pressurized combustion gas can be economically cleaned sufficiently to permit expansion through a gas turbine.

When burning high moisture lignites in the cyclone burner, a predrying system is used to maintain the temperature in the cyclone high enough to insure continuous ash melting and slag tapping. Lignite from the feeder is mixed with 750°F (400°C) primary air and carried to and through the hammer mill crusher where 10 to 12% of the total moisture is removed. This "cool" mix of moist air and dried coal is carried to a cyclone-type mechanical separator where the moist air is vented into the furnace above the top row of cyclones. The dried lignite is picked up by a rotary seal feeder and fed into another stream of 750°F air. An airlift from there to the cyclones completes the fuel feed cycle.

Modern Boiler Design for Lignite Firing

Special care must be taken in designing boilers to burn the low-rank coals, especially the lignites. Although their ignition characteristics are generally good, careful design consideration must be given to the higher moisture content, lower heating value, poorer pulverization characteristics (2), and the variable and often troublesome ash contents of these coals. In 1972, there was a total of about 2,300 MW of installed lignite-fired power plant capacity in the U.S. Since then, additional units have been placed in operation and others are under construction both in North Dakota and in Texas. By 1980, the U.S. lignite-fired capacity will have increased to about 8,000 MW.

Three recent lignite-fired boilers have been selected on which some important design specifications will be presented. These boilers were selected because they include both cyclone and pc firing, and they are designed to burn high fouling coals from North Dakota or Texas. The two 550 MW Coal Creek boilers (Figure 5.1) described in Table 5.1 are designed to burn high fouling (high sodium) lignite from McLean County in North Dakota. The 450 MW Leland Olds cyclone furnace-fired boiler No. 2 (Figure 5.2) is designed to burn high fouling lignite from Mercer County in North Dakota. Fuel specifications are shown in Table 5.2. Because of the expected fouling tendencies of these coals, the boilers were designed with considerable furnace height, 205 ft (62.5 m), low volume heat release rate, 7,200 Btu/ft^3-hr (64,074 kcal/m^3-hr) for pc firing and 12,100 Btu/ft^3-hr (107,681 kcal/m^3-hr) for cyclone furnace firing.

Figure 5.1: Cross-Sectional View of 550-MW Pulverized-Coal-Fired Coal Creek Plant, Underwood, North Dakota

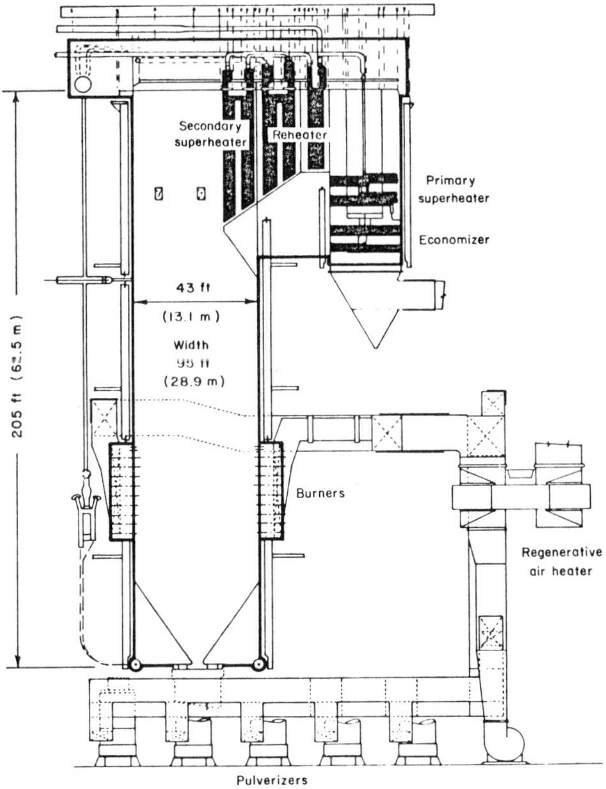

Source: "Technology and Use of Low-Rank Coals in the U.S.A.," April, 1976.

Table 5.1: Boiler and Equipment Specifications for Some Recent Lignite-Fired Powerplants in the U.S.

Plant name	San Miguel		Coal Creek		Leland Olds No. 2	
Plant rating, MW	400		Two at 550 MW (total 1,100 MW)		450	
Boiler specifications						
Manufacturer	Babcock and Wilcox		Combustion Engineering		Babcock and Wilcox	
Firing method	pulverized coal		pulverized coal		cyclone furnace	
Rated capacity, lb/hr	$3,054 \times 10^3$		$3,730 \times 10^3$		$3,250 \times 10^3$	
Pressure, psig	2,620		2,620		2,400	
Steam temperature, °F	1005		1005		1005	
Reheat temperature, °F	1005		1005		1005	
Balanced draft furnace	yes		yes		yes	
Heat release at rated capacity						
Volume heat release, Btu/hr/ft³	7,500		7,200		12,100	
Area heat release, Btu/hr/ft²	85,000		47,500		78,000	
Furnace height, ft	220		205		204	
Furnace width, ft	67		95		48	
Furnace depth, ft	45		43		45	
	Design Data					
	Flue Gas Temp (°F)	**Tube Spacing Center-to-Center (inches)**	**Flue Gas Temp (°F)**	**Tube Spacing Center-to-Center (inches)**	**Flue Gas Temp (°F)**	**Tube Spacing Center-to-Center (inches)**
Entering secondary superheater	1850	24	NA*	27	1950	22
Entering reheater	1500	9	NA	13.5	NA	9
Entering primary superheater	1110	4.5	NA	4.5	NA	4.5
Entering economizer	825	4.5	NA	4.5	NA	4.5
Soot blowers						
Type	steam		steam		steam	
Number of short wall blowers	76		192		124	
Number of long retractable blowers	80		60		157	
Pulverizer, burner data						
Pulverizer model	Babcock and Wilcox-MPS 89		Combustion Engineering-1043 RP		none	
Capacity, tons/hr	86		69		–	
Number per boiler	7		8		12 to 10 ft diameter	
Number of burners per boiler	42		64		–	

*Data not available.

Source: "Technology, and Use of Low-Rank Coals in the U.S.A." April 1976

Table 5.2: Fuel Specifications for Some Recent Lignite-Fired Powerplants in the United States

Plant name	San Miguel	
Startup date	1979	
Plant location	Christine, Texas—Atascosa County	
Company	Brazos Electric Power Cooperative Inc. and Southern Texas Electric Cooperative Inc.	
Plant rating, MW	400	
Fuel specifications		
Rank of coal	Lignite	
Location of mine	Atascosa and McMullan Counties, Texas	
	Design Analysis	**Range of Analysis**
Proximate analysis, %		
Moisture	30.0	27.0-35.0
Volatile matter	23.2	21.4-26.2
Fixed carbon	18.4	18.3-18.6
Ash	28.4	24.0-29.0
Higher heating value, Btu/lb	5,000	4,200-6,350
Ultimate analysis, % dry basis		
Carbon	41.5	40.3-45.4
Hydrogen	3.4	3.3-3.7
Nitrogen	0.73	0.68-0.84
Chlorine	0.08	0.04-0.14
Sulfur	2.39	2.2-2.7
Ash	40.6	34.9-41.6
Oxygen	11.3	10.7-12.6
	. . . . .Percent of	. . . .
	As-Mined Coal	**Total Sulfur**
Sulfur forms in coal		
Pyrite	0.45	26.9
Sulfate	0.03	1.8
Organic	1.19	71.3
Total	1.67	100.0
Grindability of coal, Hardgrove index	102	
	Design Analysis	**Range of Analysis**
Elemental analysis of ash, %		
SiO_2	62.9	61.1-65.1
Al_2O_3	17.5	16.0-19.5
Fe_2O_3	2.8	2.0-3.9
TiO_2	0.8	0.7-0.9
P_2O_5	0.06	0.04-0.12
CaO	4.8	4.0-5.5
MgO	0.7	0.5-0.8
Na_2O	3.1	2.8-3.7
K_2O	1.9	1.8-2.1
SO_3	4.6	3.3-5.9
Ash fusion temperature, °F		
Reducing atmosphere		
Initial deformation	2,120	2,070-2,170
Softening	2,386	2,360-2,400
Hemispherical	2,430	2,400-2,450
Fluid	2,700+	2,700+
Oxidizing atmosphere		
Initial deformation	2,190	2,120-2,220
Softening	2,430	2,400-2,450
Hemispherical	2,480	2,450-2,500
Fluid	2,700+	2,700+

(continued)

Table 5.2: (continued)

Plant name	Coal Creek	
Startup date	1978	
Plant location	Underwood, North Dakota	
Company	Cooperative Power Association-United Power Association	
Plant rating, MW	Two at 550 MW (total 1,100 MW)	
Fuel specifications		
Rank of coal	Lignite	
Location of mine	McLean County, North Dakota	
	Design Analysis	**Range of Analysis**
Proximate analysis, %		
Moisture	39.9	27.8-52.6
Volatile matter	27.6	21.4-32.9
Fixed carbon	25.4	11.4-31.5
Ash	7.1	3.9-11.6
Higher heating value, Btu/lb	6,258	3,068-7,660
Ultimate analysis, % dry basis		
Carbon	61.4	43.2-66.1
Hydrogen	4.6	3.1-5.2
Nitrogen	1.1	0.8-1.5
Chlorine	0.03	0.0-0.14
Sulfur	1.03	0.4-2.4
Ash	11.9	6.5-27.1
Oxygen	19.7	16.5-25.8
	..........Percent of..........	
	As-Mined Coal	**Total Sulfur**
Sulfur forms in coal		
Pyrite	0.27	26.1
Sulfate	0.04	3.8
Organic	0.72	69.9
Total	1.03	100.0
Grindability of coal, Hardgrove index	36	
	Design Analysis	**Range of Analysis**
Elemental analysis of ash, %		
SiO_2	26.5	4.7-74.6
Al_2O_3	12.4	2.2-19.9
Fe_2O_3	8.3	2.7-20.3
TiO_2	0.5	0.04-1.09
P_2O_5	0.2	0.01-0.94
CaO	24.6	8.8-42.8
MgO	6.8	2.7-12.0
Na_2O	1.3	0.1-7.3
K_2O	0.73	0.1-2.4
SO_3	17.9	3.8-32.6
Ash fusion temperature, °F		
Reducing atmosphere		
Initial deformation	2,183	1,900-2,820
Softening	2,211	1,940-2,820
Hemispherical	2,237	1,960-2,820
Fluid	2,263	1,990-2,820
Oxidizing atmosphere		
Initial deformation	2,247	2,000-2,820
Softening	2,271	2,020-2,820
Hemispherical	2,296	2,050-2,820
Fluid	2,323	2,080-2,820

(continued)

Table 5.2: (continued)

Plant name	Leland Olds No. 2	
Startup date	1975	
Plant location	Stanton, North Dakota	
Company	Basin Electric Power Cooperative	
Plant rating, MW	450	
Fuel specifications		
Rank of coal	Lignite	
Location of mine	Mercer County, North Dakota	
	Design Analysis	**Range of Analysis**
Proximate analysis, %		
Moisture	39.6	33.6-40.0
Volatile matter	27.5	–
Fixed carbon	26.6	–
Ash	6.3	4.3-9.5
Higher heating value, Btu/lb	6,523	5,613-7,320
Ultimate analysis, % dry basis		
Carbon	63.5	57.1-66.2
Hydrogen	4.5	3.9-5.3
Nitrogen	1.1	0.88-1.4
Chlorine	0.02	0.0-0.04
Sulfur	0.7	0.55-2.54
Ash	10.8	7.2-15.8
Oxygen	19.4	15.3-27.4
	...........Percent of	
	As-Mined Coal	**Total Sulfur**
Sulfur forms in coal		
Pyrite	NA	NA
Sulfate	NA	NA
Organic	NA	NA
Grindability of coal, Hardgrove index	25	
	Design Analysis	**Range of Analysis**
Elemental analysis of ash, %		
SiO_2	23.1	10.8-39.6
Al_2O_3	11.3	8.1-17.1
Fe_2O_3	8.5	4.0-15.7
TiO_2	0.45	0.26-0.68
P_2O_5	0.23	0.08-0.45
CaO	23.8	14.1-41.3
MgO	5.9	3.1-9.2
Na_2O	7.4	1.0-15.4
K_2O	0.7	0.2-1.4
SO_3	17.7	11.2-27.8
Ash fusion temperature, °F		
Reducing atmosphere		
Initial deformation	2,077	1,910-2,360
Softening	2,139	1,950-2,400
Hemispherical	–	–
Fluid	2,253	2,060-2,560
Oxidizing atmosphere		
Initial deformation	2,159	1,980-2,380
Softening	2,277	2,040-2,400
Hemispherical	–	–
Fluid	2,364	2,180-2,680

Source: "Technology and Use of Low-Rank Coals in the U.S.A.," April 1976

Figure 5.2: Cross-Sectional View of 450-MW Cyclone-Furnace-Fired Leland Olds No. 2 Plant, Stanton, North Dakota

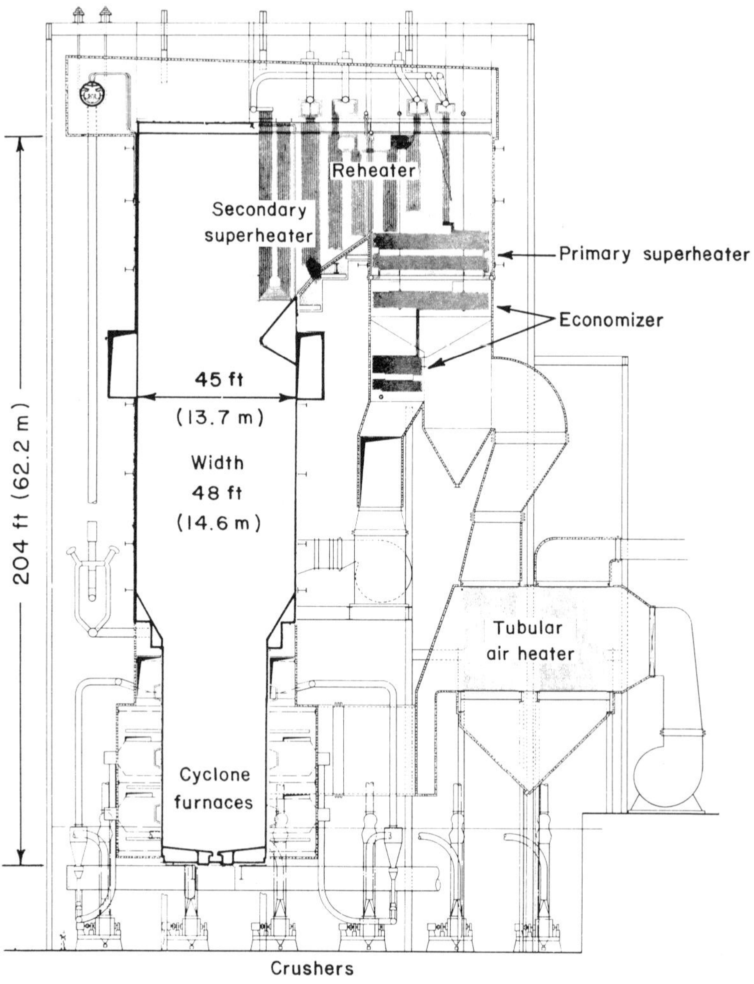

Source: "Technology and Use of Low-Rank Coals in the U.S.A.," April, 1976.

Other special features for controlling ash fouling include wide tube spacing and shallow tube bank depths in the convection section, steeply sloped floors under the pendant surfaces for shedding deposits into the main furnace, and a large number of soot blowers. The Coal Creek plant will have a total of 252 soot blowers and 27-inch (68.6 cm) spacing on the secondary superheater pendants. The Leland Olds boiler is designed to have a flue gas temperature entering the secondary superheater of 1950°F (1066°C) at rated load. Fouling in the convection section is greatly influenced by flue gas temperature as the ash fusion temperature is approached.

Figure 5.3: Cross-Sectional View of 400-MW Pulverized-Coal-Fired San Miguel Plant, Brazos Electric Power Cooperative, Inc., Christine, Texas

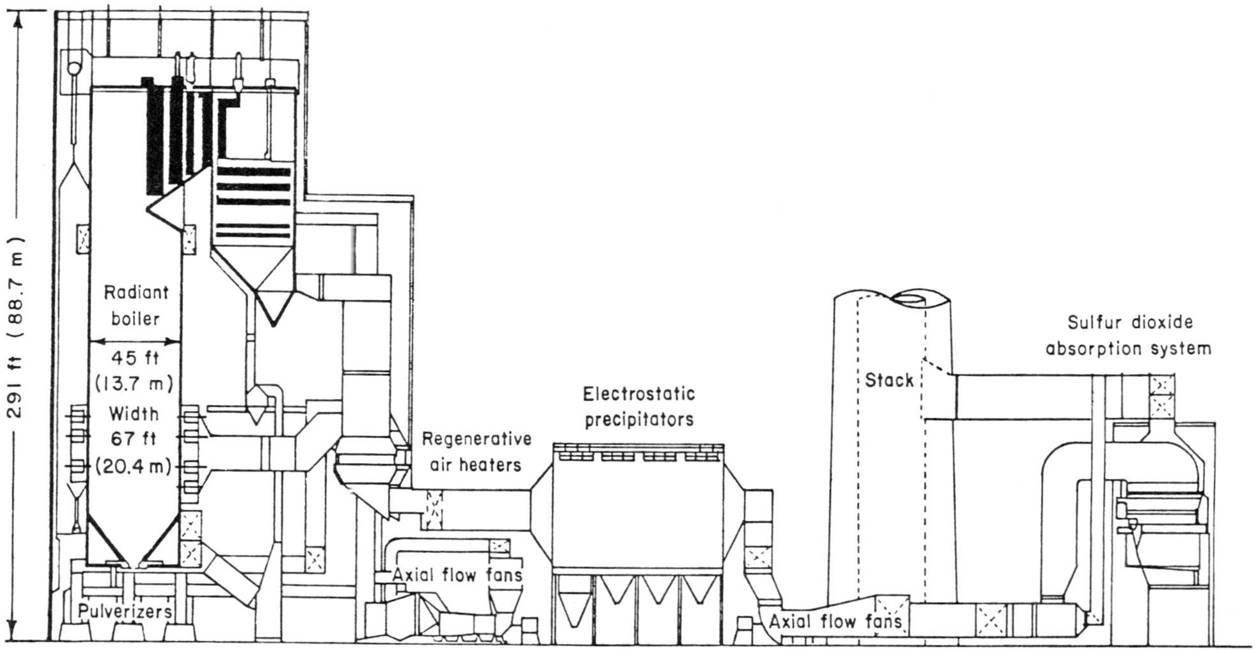

Source: "Technology and Use of Low-Rank Coals in the U.S.A.," April, 1976.

The San Miguel plant (Figure 5.3) is designed to burn a very low grade of Texas lignite (Table 5.2) having high ash (28%), low heating value (HHV, 5,000 Btu per lb, 2,780 kcal/kg), high sulfur content (2.39%) and high alkali metal content in the ash (5.0%). This is probably the lowest quality coal that has been considered for a large power station in the U.S.A., and the operating experiences of this plant will be watched with great interest by the power industry.

Ash Fouling Research

Ash fouling of heat transfer surfaces is the most serious problem encountered in operation of power boilers fired with low-rank coals. Fouling is often erratic with excessive ash deposition occurring rapidly and forcing unscheduled shutdown. Design modifications utilized to improve availability have consisted of low volumetric heat release rates, wider tube spacing, and installation of numerous soot blowers.

The Grand Forks Energy Research Center has for a number of years conducted research on properties of lignite ash and ash fouling problems. The program has consisted of testing full-scale boilers and an extensive study at the Research Center using a specially designed test furnace. The furnace has a refractory-lined combustion chamber 30 inches in diameter and 8 ft in height.

Pulverized coal is fired from a bottom burner and the fly ash laden flue gases pass through three simulated tube bank sections. Compressed air is used to maintain tube metal temperatures at superheater and reheater levels. Flue gas temperatures entering the various probe banks are about 2000°F (1093°C), 1750°F (954°C) and 1600°F (871°C). A standard test procedure has been adopted in which coal is fired for 5¼ hours, after which the test is stopped. After cooling, the weight of ash deposited on the various probe banks is determined. As a result of a test program burning various North Dakota lignites, the following criteria for fouling potential was adopted:

Deposit Weight on First Probe Bank	Relative Fouling Potential
Below 150 grams	low
150 to 300 grams	medium
Above 300 grams	high

Test furnace results have been verified by field tests on full-scale boilers. The study (3) burning various North Dakota lignites showed that by far the most important factor affecting ash deposition was the sodium content of the lignite.

The 75 lb/hr test furnace is often used to evaluate fouling potentials of coal deposits from new mining areas in advance of designing commercial boilers. For example, the Texas lignite, which will be used in the San Miguel plant, was fired in the test furnace and, incidentally, was found to be the highest fouling coal ever tested in the unit.

Additional work (4) at the GFERC is directed at determination of the chemical and physical mechanism responsible for the ash deposition phenomenon. This includes studies of the mineral matter in lignite by such methods as x-ray diffraction and electron microprobe. It is hoped that eventually some method will be found to modify the effect of the sodium so as to reduce fouling. The sodium in the low-rank coals of the Northern Great Plains exists as part of the inherent ash in the coal, and the only known method of effective removal is by ion exchange (5), which is expected to be too expensive for commercial use.

In addition to conservative boiler design, several other approaches have been used to reduce fouling problems in lignite-fired boilers. At some plants, blending of the lignites so as to control sodium content is practiced, and the high sodium lignite is scheduled to be burned during periods of lower loads. In other plants, steam soot blowers on the furnace walls have been converted to using water. This method has proved to be effective in removal of accumulated ash and slag deposits from furnace walls; however, there is some concern about the thermal cyclic stresses in the tubes which result from this operation. Water jet lancing is not thought to be suitable for removing deposits in the superheater and reheater areas.

COAL CREEK STATION STEAM GENERATORS

Introduction

The Coal Creek Station is a two-unit, 1,100-MW, mine-mouth plant being constructed in the lignite fields of central North Dakota by Cooperative Power Association of Minneapolis, Minnesota and United Power Association of Elk River, Minnesota. Both owners are generation and transmission cooperatives. Cooperative Power Association is owned and controlled by 19 rural electric distribution cooperatives in central and southern Minnesota and has over 128,000 metered accounts. United Power Association is owned and controlled by 15 rural electric distribution cooperatives in central and northern Minnesota and northwestern Wisconsin and has over 156,000 metered accounts. Cooperative Power Association and United Power Association together serve approximately two-thirds of Minnesota geographically and one-fourth of the population.

While Cooperative Power Association (CPA) and United Power Association (UPA) are separate organizations, both physically and financially, they have combined their efforts in a partnership agreement for construction of the Coal Creek Station.

Coal Creek Station is located on a 2,560-acre site approximately 6 miles south of Underwood and 50 miles north northwest of Bismarck, North Dakota. Lignite reserves, sufficient for the life of Coal Creek Station, lie within 10 miles to the north and east.

Coal Creek Station Features

The turbines are General Electric tandem compound four flow reheat units developing 550 MW at steam conditions of 2,520 psig and 1000°F with reheat to 1000°F. The generator rotor is hydrogen-cooled. Conductors in the stator are water-cooled.

The steam generators are Combustion Engineering Controlled Circulation boilers. They are each rated at 3,736,961 lb/hr at 2,620 psig and 1005°F with 1005°F reheat. The steam generators will burn pulverized lignite.

Water: The North Dakota State Water Commission has issued a conditional water permit to withdraw a maximum of 15,000 acre-feet annually from the Missouri River at a maximum rate of 14,300 gpm. A 30-inch water line will be constructed to carry the water 7 miles from the river to a 35-acre river water storage basin. The basin capacity is 860 acre-feet. Water discharge from the site is zero. Wastewater will be transferred to one of two evaporation ponds being built about 2 miles north of the main generating complex. These ponds will have a total surface area of 114 acres and contain 1,080 acre-feet. Water for mine dust control will be taken from these evaporation ponds.

Cooling Towers: Circulating water will be cooled by three round, mechanical-draft, cross flow cooling towers. Each tower is designed to supply one-third of the station cooling requirements. In addition, each cooling tower can operate with 150% of design flow, permitting operation with only two towers in service. The towers are 225 ft in diameter and 42 ft high. Each tower will have 8 two-speed reversible fans 28 ft in diameter and each fan will have 8 blades.

All three cooling towers discharge into a common 9-acre, 190-acre-foot extended basin from which the circulating water is pumped to the condensers. Use of this basin adds versatility to the cooling tower operation. Also, a cooling tower bypass pipeline permits plant start-up in cold weather without use of any cooling towers, thereby avoiding icing problems.

Circulating Water System: The combined circulating water system for both units includes four circulating water pumps, two dual pressure condensers, three round cooling towers, a cooling tower extended basin and interconnecting piping.

Four one-quarter capacity circulating water pumps of the horizontal, split case, centrifugal-type are located in the circulating water pump house and take suction from the extended basin through stationary mesh screens.

Makeup water to the circulating water system will be raw river water. Blowdown will be used for makeup to the SO_2 scrubbers and ash sluice system.

Auxiliary Cooling Water System: The auxiliary cooling water system supplies water from the cooling towers to miscellaneous cooling water heat exchangers as well as to coolers for turbine lube oil, generator hydrogen, and the excitors. Four one-third capacity auxiliary cooling water pumps of the horizontal, split case, centrifugal-type, are located in the circulating water pump house and take suction from the extended basin. The water returns to the cooling towers through the circulating water return line.

Closed Cooling Water System: The closed cooling water system will use treated condensate, cooled to 95° to 105°F by exchanging heat with auxiliary cooling water, to remove heat from the turbine controls, the air compressors, and miscellaneous lube oil and other coolers. This system is designed to dissipate 7.5 million Btu/hr. The two full capacity centrifugal closed cooling water pumps take suction from the system return header. The water is pumped through one of the two full capacity auxiliary cooling water heat exchangers and then to the equipment served. The auxiliary cooling water heat exchangers are of the horizontal, single pass, shell and tube-type. Treated condensate for the closed cooling water flows on the shell side and auxiliary cooling water on the tube side.

6.9-kV Auxiliary Electric System: The 6.9-kV auxiliary electric system supplies power to the plant auxiliaries during start-up, operation, and shutdown of the generating unit. During normal unit operation, power is supplied from the two main auxiliary transformers to serve all unit auxiliary loads. These transformers are connected directly to the generator by isolated phase bus duct. Auxiliary power required during start-up and shutdown of the unit is supplied by the one reserve auxiliary transformer provided for both generating units. Auxiliary loads are distributed among the switchgear buses so that the loss of any one bus will allow unit operation at reduced load. Either the two main auxiliary transformers or the reserve auxiliary transformer has sufficient capacity to operate the unit continuously at peak capability.

Compressed Air: Three centrifugal air compressors deliver oil-free air at 125 psig for the station requirements. One compressor normally supplies all air while a second compressor provides additional capacity during high load conditions. The third compressor is a standby unit. Upon loss of system pressure, the nonessential air requirements are automatically isolated from the compressed air system and all available air is supplied to the essential air system.

Two 500-ft^3 air receivers, three 500-scfm refrigeration air dryers, and two 300-scfm dual tower desiccant air dryers are provided. The dew point will be lowered to 35°F at 125 psig by the refrigeration dryers and to –40°F at 125 psig by the desiccant dryers. Air will be processed as required for each application to avoid ice formation or other operating problems.

Corrosion Protection and Nitrogen Supply: Nitrogen for corrosion protection is available to provide inert blanketing of the main and auxiliary steam generators and heaters when they are out of service. Other equipment will be inerted as required. The nitrogen system is supplied from a bank of nine cylinders arranged in three horizontal layers of three cylinders abreast. The cylinders are 2 ft in diameter and 20 ft long and store the nitrogen at 2,400 psi.

Auxiliary Boilers: Two lignite-fired auxiliary boilers will be provided. Each of these chain grate spreader stoker units is designed to supply 130,000 lb of steam per hr at 165 psig and about 600°F at rated capacity. These steam conditions meet the cold start-up auxiliary steam requirements for a single main unit. In addition, the auxiliary boilers are equipped with superheater elements for enthalpy matching and can be used, if required, as a steam source for the turbine-driven main boiler feed pumps. These boilers will also be used to supply auxiliary steam during low main unit load when the pressure in the auxiliary steam system drops below 150 psig and to supply building heating when both main units are out of service.

The auxiliary boilers can burn No. 6 or No. 2 fuel oil as well as lignite and they are equipped with heating coils to provide fast start-up capability.

Boiler Feed Pumps: The two one-half capacity boiler feed pumps are driven by steam turbines using extraction steam from the main turbine. With one pump out of service, the generating unit will be capable of about 60% capacity. A motor driven start-up boiler feed pump has capacity to carry the load to approximately 20% at a turbine throttle pressure of 1,200 psig.

Boiler Feedwater System: The feedwater will be heated by two high pressure heaters using turbine extraction steam. The two boiler feed pumps and start-up pump will take suction from the deaerator storage tank.

Condensate System: The condensate pumping system will take suction from the low pressure condenser and provide condensate to the deaerating heater through the gland condenser, drain cooler, and four low pressure heaters. The system will also supply condensate for miscellaneous uses such as the turbine exhaust hood sprays, the gland steam desuperheater and spray nozzles, the main steam drain desuperheater, the boiler feed pump seals, and initial filling of the boiler circulating pumps. Three one-half capacity condensate pumps are provided and are designed to provide the additional head required for a future condensate polisher if required.

Fuel Oil: The fuel oil system supplies No. 6 fuel oil to the warm-up and igniter systems on the steam generators and to the main oil burners on the auxiliary boilers. Two full capacity Dowtherm J heat exchangers heat the oil as it flows from the storage tank to the pump suction header. Oil for the warm-up burners and igniters of each steam generator is supplied by two full capacity pumps. Oil for the igniters is further heated in two full capacity steam heat exchangers to 230°F. Oil for the auxiliary boilers is also heated to 230°F in two full capacity steam heat exchangers.

Air Preheating: A two-stage air preheating system is provided to heat the combustion air. The first stage uses hot Dowtherm J in finned tube heat exchangers to raise the airflow entering the fan room to 70°F. The forced draft fan air discharge, or secondary combustion air, is further heated to control the cold end average temperature of the regenerative air heaters. Condensate from the deaerator storage section is used as a heating medium for the secondary air heating system.

Generator Cooling and Purging: Hydrogen for generator cooling will be supplied by a hydrogen generator with storage facilities. Carbon dioxide for purging the generator will come from a six-ton storage unit, which has an integral refrigeration system to keep the temperature at 0°F and 300 psi. It has an electric vaporizer capable of supplying carbon dioxide at a rate that permits purging the generator in 30 minutes.

Demineralizer: The demineralization system is a six-bed, single line demineralizer with dual primary and secondary exchangers and a two-stage vacuum degasifier between the primary and secondary exchanger pairs. The primary exchanger pairs consist of a strong acid cation exchanger and a weak base anion exchanger. Both secondary exchanger pairs and the polishing exchanger pair use strong acid cation exchangers and strong base anion exchangers. The system has a capacity of 650 gpm. Demineralized water will be stored in two 325,000-gallon tanks.

Water Pretreatment System: Service water will be supplied from a lime-softening pretreatment system. The system consists of a solids contact unit, gravity filters, a clearwell, a pretreatment waste sump, and pretreatment chemical feed. The chemicals used are pebble quick lime, alum, and a coagulant aid. The solids contact unit is sized for 1,050 gpm with a rise rate of 1.25 gpm/ft^2.

Chemical Cleaning: The chemical cleaning system includes permanent and temporary piping required for cleaning the steam generator and preboiler cycle. The preboiler cycle includes the condensate, boiler feedwater, air preheating, building heating water piping systems, and feedwater heater shells. Main headers for cleaning the steam generators and auxiliary boilers will be permanent piping. Final connection pieces will be removable. Piping for cleaning the preboiler system will be temporary.

The condensate systems, feedwater systems, air preheating, and building heating water systems will be flushed with demineralized water. The feedwater heater shells will be soaked with hot demineralized water. The preboiler systems and steam generators will be cleaned with conventional acid and alkaline solutions.

Fire Protection: The fire protection system has been designed to provide protection for the yard structures, generation building, scrubber, and coal handling facilities.

Potable quality water will be stored in the water storage tanks located adjacent to the water treatment building. Water will be supplied to the tanks from the water treatment system at a rate of 1,000 gpm. Fire and service water is supplied to the station by a common system which includes two full capacity service water pumps, a motor driven fire pump, and a diesel engine driven fire pump.

Fire hydrants are spaced about 300 ft apart along the ring header around the perimeter of the generation area. Hydrant houses are provided at various locations throughout the plant site and are provided with fire hoses, fire axes, fog nozzles, hydrant wrenches, and spanners. Separate booster pumps are provided where necessary for higher elevations around the steam generators and lignite conveyors.

Water spray systems are provided for protection of storage and warehouse facilities, turbine lubricating oil storage, main turbine oil supply, drain piping, hydrogen seal oil units, generator transformer, and main and reserve auxiliary transformers. Carbon dioxide gas supplied from a central liquid bulk storage unit will be available at hose cabinets in the central control bay, at grade, mezzanine, and operating floor levels for use if required at switch gear, relay, and control rooms. Portable extinguishers will be provided as required for general fire protection. Individual carbon dioxide systems are provided for protection of the lignite dust collectors.

The lignite pulverizers are protected by both an individual carbon dioxide inerting system and a steam blanketing system.

Ashes: The bottom ash and scrubber sludge will be pumped to a common disposal area through a common line. Since the fly ash is approximately 25% lime by weight, as much of it as possible will be used as an additive to the scrubber.

The ash disposal area will cover 105 acres and has a storage capacity of 3,400 acre-ft. The area can be expanded to accommodate waste for the life of the station.

Fuel Supply: Independent studies by Cooperative Power Association and United Power Association showed that construction of a large mine-mouth generating station in North Dakota was the most economical solution to their long-term power needs. Property adjacent to the southern edge of North American Coal Company's Underwood field was selected as the site for the Coal Creek Station.

Significant locations are indicated on Figure 5.4, a map of the area. The Missouri River is some 7 miles south of the site, and U.S. Highway 83 and tracks of the Soo Line Railroad are less than 2 miles east of the site.

Figure 5.4: Map of Coal Creek Area

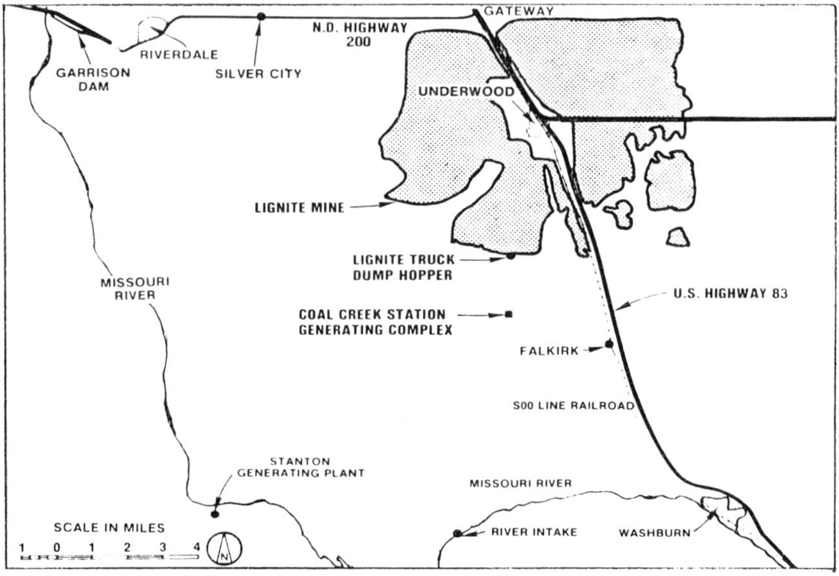

Source: NTIS GFERC/IC-77/1

Although the exploration and leasing program carried out by North American Coal Company included drilling and analysis to determine the quantity, quality, and mineability of the Underwood reserves, the data was not considered adequate to serve as a basis of design parameters for the fuel preparation equipment, steam generator, combustion gas cleaning equipment, and ash handling equipment.

Because large variations can occur in lignite quality within a given seam, an extensive sampling and analysis program involving about 200 core samples was undertaken.

The lignite reserves consist of two seams or benches separated by parting. Each bench was sampled independently along with the parting. The location of the mine was defined (the shaded area in Figure 5.4) and mean fuel characteristics and properties were derived from analytical data for cores taken from the area that will be mined. Mean, minimum, and maximum values determined in the sampling program are listed for the lignite data in Table 5.3 and for the ash data in Table 5.4. Appreciable variations in fuel quality will be encountered. It is anticipated that, during the course of normal mining operations, blending of the various qualities of lignite will occur so that the fuel supplied will not exceed the reject limits specified in the coal sales agreement. Limits of principal properties are:

	Reject Limit	. . . .Falkirk Reserves	
		Mean	Range
Heating value, Btu/lb	6,000	6,258	3,068-7,660
Moisture, %	42	39.83	27.78-52.55
Ash, %	9	7.14	3.89-15.95
Sulfur, %	1.4	0.63	0.18-1.41

During the first several months of station operation, lignite of average quality or better will be burned rather than lignite of minimum quality or better. As experience is gained in operation of the station, lower quality lignite can be used.

Table 5.3: Lignite Analysis

	Mean	Range
Proximate analysis as received, % by wt		
Moisture	39.83	27.78-52.55
Ash	7.14	3.89-15.95
Volatile matter	27.61	21.40-32.93
Fixed carbon	25.42	11.39-31.50
Total	100.00	
Sulfur	0.63	0.18-1.41
Heating value, Btu/lb	6,258	3,068-7,660
Ultimate analysis as received, % by wt		
H_2O	39.83	27.78-52.55
Ash	7.14	3.89-15.95
Sulfur	0.63	0.18-1.41
Nitrogen	0.65	0.52-0.95
Carbon	37.21	21.16-39.87
Hydrogen	2.78	1.52-3.42
Oxygen	11.74	10.71-12.45
Chlorine	0.02	0.00-0.08
Total	100.00	
Sulfur forms as received, % by wt		
Pyritic	0.17	0.02-0.82
Sulfate	0.02	0.00-0.26
Organic	0.44	0.05-0.93
Total average	0.63	0.18-1.41
Equilibrium moisture, % by wt	35.51	24.80-49.30
Alkalies as Na_2O as received, % by wt	0.08	0.002-0.397
Silica value	39.5	8.0-83.0
Hardgrove grindability index	35.90	22.00-136.00

Source: NTIS GFERC/IC-77/1

Table 5.4: Ash Analysis

	Mean	Range
Ash analysis, % by wt		
Phosphorus pentoxide, P_2O_5	0.16	0.01-0.94
Silica, SiO_2	26.49	4.66-74.55
Ferric oxide, Fe_2O_3	8.30	2.65-20.30

(continued)

Table 5.4: (continued)

	Mean	Range
Alumina, Al_2O_3	12.43	2.22-19.90
Titania, TiO_2	0.49	0.04-1.09
Lime, CaO	24.61	8.80-42.80
Magnesia, MgO	6.83	2.70-12.00
Potassium oxide, K_2O	0.73	0.11-2.36
Sodium oxide, Na_2O	1.25	0.12-7.25
Sulfur trioxide, SO_3	17.93	3.77-32.62
Undetermined	0.78	0.01-1.09
Total	100.00	
Ash fusion temperatures, reducing, °F		
Initial	2183	1900-2820
Softening	2211	1940-2820
Hemispherical	2237	1960-2820
Fluid	2263	1990-2820
Ash fusion temperatures, oxidizing, °F		
Initial	2247	2000-2820
Softening	2271	2020-2820
Hemispherical	2296	2050-2820
Fluid	2323	2080-2820

Source: NTIS GFERC/IC-77/1

Transmission Lines: The majority of the energy supplied by the Coal Creek Station will be converted to direct current in a thyristor-based converter station designed and supplied by ASEA of Sweden. The output of the converter will be ±400 kV dc. The energy transmission on the dc line is from North Dakota to Minnesota only. The transmission line will be bipolar, two conductors per pole: 1,590-kcm ACSR 45/7 Lapwing. The transmission line towers will be self-supporting, galvanized, lattice-steel structures spaced an average of 1,400 ft over the 430-mile route from the Coal Creek Station to a substation near Delano in Wright County, Minnesota. Conductor ground clearance will be a minimum of 35 ft with a 50-ft minimum over irrigated land. A straight line structure will have a base 35 feet square.

Two ac transmission lines will also be constructed in North Dakota. One side of a double 230-kV line will tie to an existing UPA transmission line about 5 miles west of Coal Creek Station. The other side will run adjacent to this existing line an additional 7.5 miles to the UPA's Stanton generating plant. These 230-kV lines will provide start-up power as well as a tie to the North Dakota power grid. The second ac transmission line will be a 69-kV line to supply power for the lignite mining operation.

Two 345-kV ac transmission lines will be constructed for transmission and distribution of the electricity to Minnesota load points from the substation near Delano. A single-circuit line, approximately 76 miles long, will tie into the Minnesota power grid at a point near Mankato. A double-circuit line, 28 miles long, will tie into the Minnesota power grid north of Minneapolis.

Lignite Handling System

General Description: Lignite is supplied to the Coal Creek Station from the Falkirk Mine operated by the Falkirk Mining Company, a wholly-owned subsidiary of the North American Coal Corporation.

The lignite handling system will convey fuel from the mine to the steam generators of Units 1 and 2 and also to the two lignite-fired auxiliary boilers. The maximum usage rate is expected to be 800 tph when both units are operating at maximum capacity. The fuel supply to the generating units will be highly reliable because of the alternatives available. One million tons, a 60-day supply, will be maintained in inactive storage which can be reclaimed by mobile equipment through either the stacker-reclaimer or the emergency reclaim hopper system.

In addition, the system will have active storage of 126,000 tons, recoverable by the stacker-reclaimer, and live storage of up to 48,000 tons in three storage silos. This live storage will provide the station requirements for a typical three-day weekend of operation. The conveying system between the reclaim facilities and the steam generators is completely redundant. Flexibility is provided in the design of the system to permit automatic blending of fuel from active storage with lignite supplied from the mine.

Conveyors: Lignite will be transported 1.4 miles from the 16,000-ton mine surge silo to the Coal Creek Station generating complex over the conveying system shown in Figure 5.5. The longest conveyor in the system, about 4,300 ft between horizontal centers, receives lignite from the six mine silo outlet feeders at rates to 2,500 tph and carries it to the transfer building. A belt scale near the mine silo and an as-received sampling system in the transfer building will be used for revenue determinations.

Figure 5.5: Lignite Conveying System

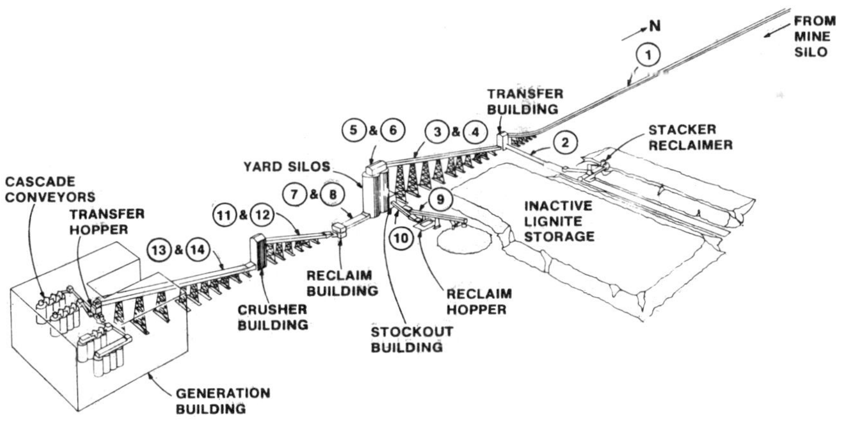

Source: NTIS GFERC/IC-77/1

Lignite entering the transfer building may be transferred to the stacker-reclaimer conveyor or onto either of the parallel, redundant conveyors to the yard silos. The lignite on either conveyor may be unloaded into either of the two 16,000-ton concrete silos by using gates and transfer conveyors. The silos are 70 ft inside diameter by 213 ft high. Space has been provided for additional silos south of the present silos for possible expansion of the system. Lignite will be removed from each silo through seven gates and feeders onto either of the two horizontal reversible conveyors operating beneath the silos. When the horizontal reversible conveyors carry fuel toward the station, they can be loaded either from the silo vibrating feeders or from inactive storage through the emergency reclaim hopper and conveyor system. Lignite may be placed in inactive storage by reversing the direction of conveyor travel beneath the silos, thereby transporting lignite to the emergency stocking-out conveyor.

In normal operation, lignite is transferred from the horizontal reversing conveyors to redundant inclined conveyors that terminate in the crusher building. The crusher building contains magnetic separators for tramp iron, a hopper, and two 1,000-tph crushers. The primary purpose of the crushers is to break frozen lumps of lignite. A crusher bypass is included for use during warm weather. A small portion of the lignite conveyed from the crushers to the transfer hopper in the generation building is diverted to the as fired sampling system, which includes cutters, a crusher, a reject elevator, and the required conveyors.

Lignite is discharged from the transfer hopper through four gates and vibrating feeders onto the conveyors, which carry fuel to the four cascade conveyors on each side of each steam generator. The cascade conveyors discharge into the steam generator silos.

Flight conveyors, under the elevated portions of the principal conveyors feeding the transfer hopper, contain dribble and carry it to the transfer hopper. Bend pulley hoppers under the four conveyors transporting fuel from the transfer hopper to the cascade conveyors, discharge dribble to bucket elevators that lift the fuel to the discharge chutes of the conveyors they serve.

Lignite for the auxiliary boiler is removed from the transfer hopper by screw conveyors and lowered about 70 ft by a flight conveyor to the auxiliary boiler day hoppers. Each day hopper can either discharge lignite to the associated auxiliary boiler or can be emptied through an unloading chute during periods of extended auxiliary boiler outage, when station steam services are supplied by the main boilers.

Stacker-Reclaimer: The stacker-reclaimer is a remotely operated machine that operates in conjunction with a 60-inch reversible yard conveying belt. The base of the machine consists of a rail-mounted traveling gantry straddling the yard belt. The gantry carries a slewing mast structure capable of rotating 328° about a vertical axis. An average stacking rate of 2,500 tph and a reclaiming rate infinitely variable from 500 to 2,000 tph can be maintained. Lignite will be stacked in windrows, each containing 70 tons per linear foot, up to 900 ft long on each side of the rails.

Belt Scales and Controls: In addition to the belt scale at the mine silo used for revenue determination, several additional scales are provided for inventory and control purposes. Lignite quantities are monitored at the following locations:

- Mine silo discharge
- Stockout conveyor
- Reclaim conveyor
- Yard silo discharge
- Crusher discharge
- Transfer hopper feeder discharge
- Each auxiliary boiler

Silo and hopper quantities are also monitored, thus permitting automatic operation of the lignite supply system. After the stacker-reclaimer operation is manually set to the initial position of either the stacking or reclaiming mode, the system can operate automatically. The station operator will monitor and operate the lignite handling system from the control room. There are local stations near the transfer hopper, the crusher building, the yard silos, the stacker-reclaimer, and the mine silo for local control of equipment as required.

Dust Control: Dust generated by transferring or crushing activities is collected by fabric dust collectors with air exhaust fans at the following locations:

- Transfer building
- Top of yard silos
- Silo outlet and emergency stockout area
- Crusher building
- Transfer hopper and auxiliary boiler hoppers
- Unit 1 cascade conveyors
- Unit 2 cascade conveyors

The dust collected is returned to the conveying system.

Low Ambient Temperatures: Low ambient temperatures were considered in the design of the lignite handling system. First, all conveyors not enclosed in the heated main plant building are equipped with a creep drive and lubricated with low temperature lubricants. The exterior conveyors will be operated continuously at either creep or regular speeds during cold weather. Second, mine and yard silos are provided to absorb surges in mine production or in reclaiming activities. In cold weather, residence time of lignite in these silos will be held to a minimum to prevent formation of frozen lignite.

Finally, with the exception of the yard conveyor, which operates in conjunction with the bucket wheel stacker-reclaimer, all exterior conveyors are covered to reduce the loss of fines and to protect the conveying equipment. The conveyor belt from the mine to the transfer building has arched covers. Inclined conveyors between the transfer building and the generation building are enclosed but unheated, with walkways to facilitate maintenance and inspection.

Unloading the Lignite Handling System: If it becomes necessary to empty the conveying system because of an unscheduled outage, lignite in the mine silo can be conveyed to the stacker-reclaimer or to the emergency stockout pile. The contents of the yard silos can be transferred to the emergency stockout pile, and lignite on the belts between the yard silos and the station, or in the crusher hopper, can be removed from the belts at a point south of the crusher building by activating plows mounted above the conveyor belts entering the station. The 120-ton auxiliary boiler day hoppers can be unloaded at ground level through unloading chutes.

The eight silos above the pulverizers in each of the two main generating units are provided with removable outlet adapters to permit removal of lignite through portable screw conveyors in the event of an extended pulverizer or unit outage.

Steam Generator Design

Steam Generators: The steam generators serving Units 1 and 2 at the Coal Creek Station are Combustion Engineering Controlled Circulation, balanced-draft, divided furnace boilers (Figure 5.1), with a primary steam flow of 3,763,961 lb/hr each at maximum continuous rating. The superheater outlet steam temperature is 1005°F at a pressure of 2,620 psig with a feedwater temperature 490°F. The multistage reheater will raise 3,343,430 lb/hr of steam to an outlet temperature of 1005°F from inlet conditions of 636°F and 610 psig. The boilers are of conventional type with a furnace deflection arch and a radiant reheater wall. The units were specified for and are designed to fire North Dakota lignite with the following analysis:

	Typical	Range
Proximate analysis, %		
Moisture	36.6	35-42
Volatile matter	26.4	24-30
Fixed carbon	30.8	26-31
Ash	6.2	4-9
Total	100.0	-
Higher heating value, Btu/lb	6,800	5,780-7,100
Grindability, Hardgrove index	35	25-45
Ash softening temperature, °F (H = W)	2188	-

Furnace Design: Each boiler has a 96-ft wide by 42-ft 11-¼-inch deep furnace, divided into two cells by a center wall. The height of the furnace is 205 ft and is void of any projecting surfaces or platens in or over the furnace proper. The outer walls are fusion welded, consisting of 1¾-inch o.d. tubes on 2¼-inch centers. The center wall is comprised of 2¼-inch o.d. tangent tubes. The tubing material is SA-210 A-1 carbon steel. The furnace combustion rate is 7,600 Btu per hour per cubic foot. The net heat release rate is 47,500 Btu per hour per square foot to the exit of the front reheater platen section. The maximum gas temperature leaving the furnace is specified at 1900°F.

Circulation System: Each boiler has a 66-inch i.d. steam drum. From the steam drum, eight 16-inch o.d. downcomers connect to the 30-inch o.d. pump suction manifold. The total circulation is distributed from this manifold to four inverted, dual-discharge submerged-motor boiler water circulating pumps. Each pump is rated at 13,200 gpm and the electric characteristics of each motor are 3-phase, 60 cycles, 1,800 rpm, 6,600 volts. The pump discharge piping feeds the two lower furnace drums which, in turn, feed the furnace wall system. Each furnace wall tube is orificed to control the amount of flow, so that the greatest flow is directed to the areas of highest heat absorption. A system of 6-inch o.d. relief tubes carry the generated steam/water mixture from the furnace wall outlet headers back to the 66-inch i.d. steam drum, completing the circuit. The saturated steam is separated from the saturated water by the steam drum internals. Saturated steam then leaves the drum through 6-inch o.d. tubes to the superheater.

Superheater, Reheater, and Economizer: Figure 5.6 illustrates the superheater, reheater and economizer.

Figure 5.6: Superheater, Reheater, and Economizer Schematic

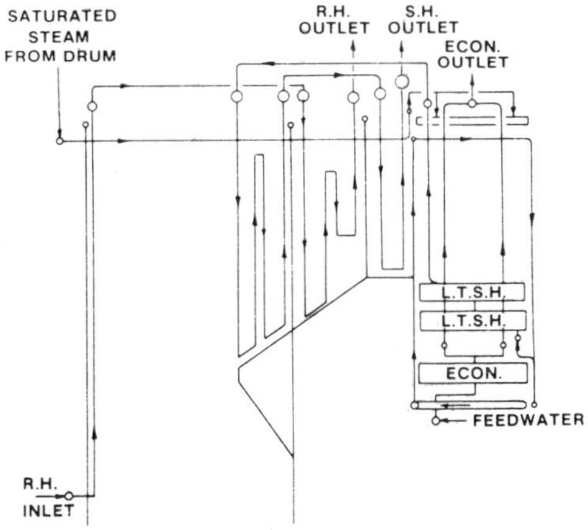

Source: NTIS GFERC/IC-77/1

The superheater circuit comprises the peg-finned steam-cooled roof, fin-welded rear pass walls, a low-temperature horizontal section, pendant platens in the furnace arch region, and a finishing section in the convection pass. Desuperheaters are located in the piping between the low temperature horizontal section and the pendant platens.

The reheater consists of a radiant section covering the front wall of the upper furnace and a finishing section located in the convection pass behind the superheater platens.

The plain tube economizer consists of one bank of 2-inch o.d. tubes on 4-½-inch transverse spacing and is located below the horizontal superheater section. Both the horizontal superheater section and the economizer are supported on the economizer outlet terminal tubes.

Steam Temperature Control: The ranges of constant 1005°F steam temperature are 1,655,000 to 3,736,961 lb/hr primary superheat steam and 1,510,000 to 3,343,430 lb/hr reheat steam. Over this load range, reheat temperature is controlled by fuel nozzle tilt and the superheat temperature is controlled by interstage desuperheating. Positioning of the fuel nozzles between 30° below to 30° above the horizontal provides a means of automatically adjusting the furnace effectiveness as required by changes in load or in furnace conditions. While design of the unit is such that no need of desuperheater water is expected for controlling reheat steam temperature, reheat desuperheaters are provided in the cold reheat piping to take care of any abnormal or emergency conditions.

Fuel Burning System: When operating at the design rating and firing the typical lignite, 793,200 lb of fuel are required per hour for each unit. The lignite will be pulverized by 8 C-E RP-1103 bowl mills per unit with 110-inch diameter bowls. When these mills enter service, they will be the largest power plant coal pulverizers in operation. The RP-1103 pulverizers are furnished with an external or separate gear case that isolates the gearing from the high mill air temperatures associated with lignites.

Also, if required, the gear case can be removed for maintenance without dismantling the pulverizer. The pulverizers grind the fuel to a fineness of 65% through 200 mesh. The pulverized fuel is delivered to the furnace through 20-inch pipes and admitted to the furnace through the tilting tangential fuel nozzles in the 8 windbox registers located in the front and rear walls of the divided furnace. Each pulverizer supplies a single elevation of fuel nozzles, one per windbox.

Each windbox contains 8 elevations of fuel nozzles with each pair of nozzles receiving ignition energy from a No. 6 oil gun located between them. Each oil gun level contains its own high energy arc ignitor. Above and below each fuel nozzle are the auxiliary (secondary) air compartments, which admit hot secondary air into the furnace from the two Ljungstrom air preheaters.

Nitrogen Oxides Control: The design criteria for NO_x emissions from the station are specified to not exceed 0.7 lb/million Btu heat input.

With tangential firing, the lateral impingement of the horizontally adjacent fuel and air streams produces a furnace vortex with a single flame envelope. The entire furnace acts as the burner; therefore, precise proportioning of fuel and air at each of the individual fuel and air admission points is not required. Locally fuel-rich or air-rich streams are mixed in passing through the furnace, resulting in complete combustion of the fuel. The furnace vortex produces a large amount of internal recirculation of bulk gas, which coupled with the longer residence time for burning, provides a combustion system inherently low in NO_x production and virtually eliminates hydrocarbon and CO emissions.

Further reduction of NO_x emissions is achieved by the use of overfire air. Overfire air compartments are provided as extensions of the windboxes. Overfire reduces NO_x formation by reducing the peak and bulk flame temperatures by extending the combustion zone and time necessary for fuel burnout. The overfire air is directed into the furnace through two elevations of separately tilted windbox registers. Optimum damper and tilt positions are established by field testing.

Air Preheaters: Each steam generator is equipped with two vertical shaft Ljungstrom trisector regenerative air preheaters. These heaters are divided into three sections: primary air (air for fuel drying and conveying), secondary air (air which goes directly to the furnace windboxes), and flue gas.

The air preheaters are equipped with three layers of heating elements designated hot, intermediate, and cold layers. They are designed to provide 768°F primary air to the pulverizers and 741°F secondary air to the windboxes, using 100°F inlet air at the maximum continuous rating of the steam generator. Exiting flue gases will be cooled from 838° to 338°F. Coils located between the forced draft fans and the air preheaters will heat the incoming air to maintain an average cold end air preheater temperature on the Ljungstrom of 215°F.

Fans: Each steam generator will have two 1,200-rpm centrifugal primary air fans to supply the air for pulverized fuel drying and transport. Air from the primary air fans passes through the primary air section of the air preheater directly to the pulverizers. The amount of hot primary air passing through this section of the air preheater is determined by the number of pulverizers in operation, pulverizer loading, and the fuel moisture. If required, cold primary air can be mixed with the hot primary air via tempering dampers to provide the required pulverizer inlet air temperature for fuel drying.

Two 900-rpm centrifugal forced draft (FD) fans will supply the combustion air through the secondary section of the air preheater directly to the windboxes. Four 900-rpm centrifugal two-speed induced draft (ID) fans will be supplied on each steam generator.

Soot Blowers: Each unit will be equipped with steam soot blowers. There will be 28 double-sided, retractable soot blowers (approximately 48 ft lance travel) located at the furnace arch, superheater, and reheater sections. Furnace wall cleaning will be accomplished with 170 wall blowers on five elevations between the top fuel nozzle level and the furnace arch, with one elevation of 34 wall blowers below the lowest fuel nozzle level. Two soot blowers, one per side, are also located just above the furnace bottom opening, for a total of 262 soot blowers per unit. A programmable controller will control the soot blowing system. The locations of the retractable and wall blowers are shown in Figure 5.7.

Figure 5.7: Location of Retracts and Wall Blowers

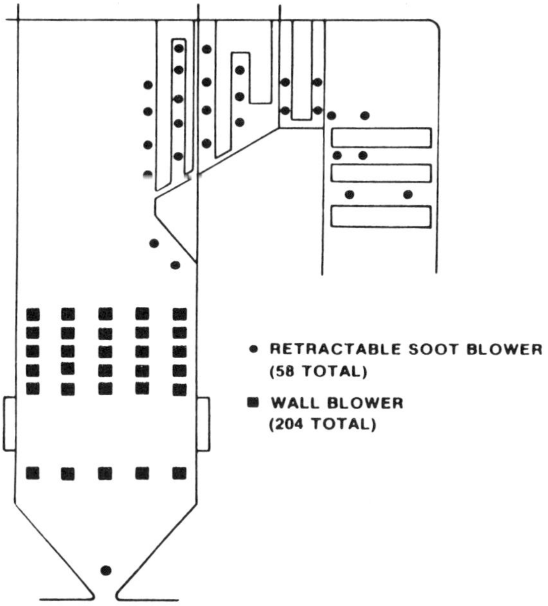

Source: NTIS GFERC/IC-77/1

Electrostatic Precipitators

Arrangement: Fly ash from the Coal Creek Station steam generators will be collected in precipitators operating at about 321°F. Each of the two generating units have a separate flue gas treatment and train stacks. Flue gas leaving each air heater is divided into four parallel gas streams that permit duct arrangements providing uniform gas distribution. Each of the four gas ducts leaving the precipitators connects to dual inlet boxes of four separate induced draft fans. Gases discharged from these fans are combined into two ducts that pass either side of the stack to enter the flue gas scrubbing systems.

Description: The electrostatic precipitator for one generating unit is shown in Figure 5.8. The enclosure (not including duct distribution boxes) is about 95 ft high, 217 ft wide, and 107 ft long in the direction of gas flow. An 8-ft clearance under the 48 fly ash hoppers provides access for removal of ash spills during maintenance. The control room and 48 transformer/rectifiers are above the gas enclosures and under the roof (Figure 5.8).

Figure 5.8: Electrostatic Precipitator

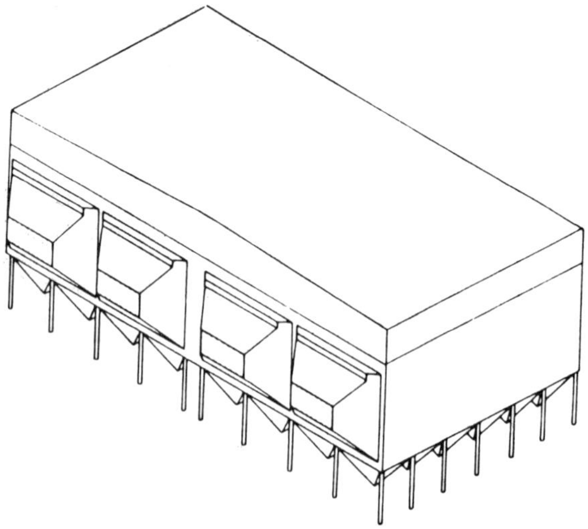

Source: NTIS GFERC/IC-77/1

The precipitator is constructed as two separate shells. The gas duct arrangement permits operation with a single shell in service at reduced unit load. Fly ash is removed from the hoppers by a pressurized pneumatic system that conveys the ash to two large yard silos and four smaller silos in the scrubber buildings. (Ash will be used as a scrubber reaction additive.) Ash from the yard silos will be conditioned by the addition of a small amount of water for dust control as it enters trucks for transportation to the disposal site. Facilities are provided to permit loading dry unconditioned fly ash into railroad cars for sale if a market develops.

Design Parameters: Gas enters the precipitator of each generating unit at the following design conditions:

Flow rate, lb/hr	6,319,000
Actual cubic feet per minute (at 321°F and 26.14" Hg abs)	2,310,000
Temperature, °F	321
Inlet dust loading, grains per actual cubic foot (at 321°F and 26.14" Hg abs)	1.1 to 3.58

The precipitators will remove 99.5% of the dry solid particulate at inlet loadings greater than 1.16 grains per actual cubic foot. At lower inlet loadings, the outlet loading will be 0.0058 grains per actual cubic foot or less.

Spray Tower Scrubbers

Design Parameters: Each steam generator will produce 6,319,000 lb of flue gas per hour when operating at maximum continuous capacity. This gas will carry from 5,000 to 14,000 lb of sulfur dioxide per hour, depending on the sulfur content and quality of the lignite being burned. Pursuant to the Clean Air Act of 1970, the U.S. Environmental Protection Agency and the State of North Dakota adopted air quality standards that designated sulfur dioxide as a pollutant. These standards limit both the amount that can be discharged and the total quantity in the ambient air after the plant discharges have been added to the previously existing background concentration. The limiting quantities applicable to Coal Creek Station are:

	Sulfur Dioxide
Emission, lb/million Btu input	1.2
Ambient air concentration, $\mu g/m^3$ (ppm)	
1-hour average	715 (0.28)
3-hour average	1,300 (0.50)
24-hour average	260 (0.10)
Annual average	60 (0.02)

At maximum capacity, the heat input to each steam generator will be 5,280 million Btu per hour. Without gas cleaning equipment, sulfur dioxide emissions could range, with varying fuel sulfur contents, from 0.9 to 2.6 lb per million Btu input. Although the sulfur content of the lignite to be used in Coal Creek Station is relatively low, the sulfur dioxide emission standards of 1.2 lb per million Btu input could not always be met by fuel blending. Some fuel will not require scrubber operation, but scrubbing of some flue gas will usually be required.

Evaluation of feasible alternatives resulted in selection of the partial scrubbing concept. This system uses high efficiency scrubbers to process a maximum of about 60% of the flue gas. The remainder of the gas will bypass the scrubber and be mixed with the scrubbed gas to reheat it. This partial scrubbing system offers several advantages over systems that require the scrubbing of all the flue gas:

- Lower costs
- Elimination of reheat equipment
- Improved station heat rate
- Lower makeup water requirements

The scrubbers are designed to remove 90% of the sulfur dioxide from up to 60% of the flue gas.

Combustion gases will enter the scrubber ductwork at about 321°F. The gases being scrubbed will leave the scrubbers at about 135°F. If, for example, 60% of the flue gases are being scrubbed and 40% bypassed, the temperature of the gas stream after recombination and mixing would be about 210°F. This is the lowest anticipated temperature for gases entering the stacks and should be adequate to provide plume buoyancy and reduce water visibility.

Configuration: The scrubbers are C-E spray tower air quality control systems (AQCS) (Figure 5.9). Flue gas discharged by the four induced draft fans merge into two ducts that carry the gas into the scrubber ductwork. The scrubber system for each generating unit includes four countercurrent spray tower modules. Any module may be isolated by closing the two inlet and two outlet louver dampers and pressurizing the space between them. Opposed vane dampers are used to equalize the gas flow between the parallel scrubber modules and the hot gas bypass duct.

The scrubber modules contain three levels of spray nozzles aimed downward. Alkaline slurry from the reaction tanks is intimately mixed with flue gas flowing upward through the modules.

Figure 5.9: Spray Tower Scrubbers

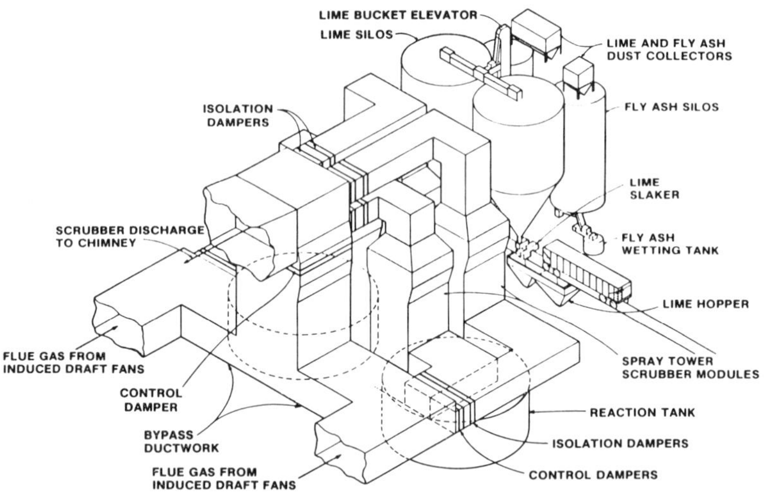

Source: NTIS GFERC/IC-77/1

Droplets are removed from the exiting gas stream by a bulk entrainment separator and two layers of mist eliminators. Gas leaving the top of the scrubber modules will be essentially free of entrained moisture, but will be saturated at about 135°F. The scrubber modules and the outlet ductwork will be constructed on Type 316L stainless steel. The stack liner is constructed of acid-proof brick.

The alkaline additive to the reaction tanks will be fly ash supplemented as required by slaked pebble lime. The location of the storage silos in the scrubber building is shown in Figure 5.9. Pebble lime will be delivered by truck or railroad hopper cars. Fly ash will be transferred to the scrubber silos by the pressurized pneumatic system serving the precipitator hoppers. Pebble lime from the silos is fed to slakers, then to an additive holding tank, from which it is pumped to the reaction tanks. Fly ash is fed to a wetting tank, then gravity-fed to the reaction tanks. Slurry blowdown from the reaction tanks is discharged to the bottom ash storage pond through the bottom ash lines.

Summary

Cooperative Power Association and United Power Association are the owners of the Coal Creek Station in central North Dakota. Two generating units with a gross output of 550 MW each have been under construction. The Coal Creek Station is a mine-mouth plant and will burn lignite supplied concurrently from two pits operated by a subsidiary of the North American Coal Company. Sufficient fuel for the life of the plant is within 10 miles of the steam generators. The fuel supplier will deliver crushed lignite to within 1.4 miles and it will then be transported the remaining distance to the boilers by conveyor belts.

The steam generators are balanced draft Combustion Engineering reheat units rated at 3,736,961 lb of steam per hour at 2,620 psig and 1005°F with 1005°F reheat. Steam temperature will be controlled with the tilting tangential firing system as well as with desuperheating. Overfire air will minimize NO_x emissions. Each steam generator uses two trisector air heaters with two primary and two secondary air fans to provide independent heating of the primary air used to dry the high moisture content lignite in the pulverizers.

The flue gas treating systems of the two generating units are independent. Gas leaving the air heaters flows through precipitators that remove 99.5% of the dry solid particulate matter. The induced draft fans are situated between the precipitators and the scrubbers. The scrubbers are designed to remove 90% of the SO_2 from up to 60% of the flue gas. The remainder of the flue gas bypasses the scrubbers and is used to reheat the scrubbed gas. Emissions from the 650-ft brick-lined stacks will be well under regulatory limits.

FLUIDIZED-BED COMBUSTION OF NORTH DAKOTA LIGNITE

Introduction

Fluidized-bed combustion (FBC) is the combustion of fuel in a bed of inert or chemically active particles which are held in suspension by an upward flow of air. The minimum air velocity required to fluidize the bed is defined by the balance between the aerodynamic drag force on the particles and the gravitational force. The earliest application of the fluidized combustion process is reported by Skinner (6) to be the Winkler Gasification Process of about 1930. At the present time, there is great interest in the fluidized combustion of coal because it provides an alternate method of burning noncompliance coal in an environmentally acceptable manner. DOE is the lead agency in the development of this technology, and is currently operating or in the process of starting up process development units, component test units, and demonstration size units, the largest of which is the 30 MW_e unit at Rivesville, West Virginia.

The role of the Grand Forks Energy Research Center in the fluidized-bed combustion program is to develop data on sulfur retention on the alkaline coal ash from Western United States coals, when burned in a fluidized-bed combustor without an additive. FBC will result in substantially lower SO_2 emissions than from conventional combustion systems and may, therefore, meet the Federal New Source Performance Standard of 1.2 lb $SO_2/10^6$ Btu. The effects of operating conditions and coal ash composition on the retention of sulfur dioxide during fluidized-bed combustion of Western coals will be evaluated to provide a design base for these coals.

The fluidized combustion of coal offers several important advantages over conventional pulverized-coal-fired furnaces. The fluid bed can be operated on much larger (usually about ¼ inch top size) and less uniform coal particle sizes. The fluidized-bed boilers will be smaller than conventional boilers because of the high heat transfer rates to the steam tubes immersed in the fluid bed. The fluid bed will operate at much lower temperatures than conventional combustion systems, with the normal range expected to be from 1300° to 1900°F. This low temperature allows a greater sulfur adsorption by the alkaline ash, and reduces the emission of the oxides of nitrogen.

The principal disadvantages of the fluidized-bed combustion process are: (1) the carry-over of unburnt coal particles because of the high gas velocities; (2) the production of relatively large amounts of carbon monoxide which lowers combustion efficiency; and (3) the corrosion and erosion problems associated with the immersion of heat transfer surfaces in the bed of fluidized particles.

The ash of lignites and Western subbituminous coals contains a high percentage of the alkali constituents, calcium, magnesium, and sodium. These coals, when burned in a conventional pc-fired furnace, will retain from 10 to 40% of the sulfur on the ash. These same coals, when burned in a fluidized-bed combustor, should retain a greater percentage of sulfur on the ash because of: (1) the increased contact time between the sulfur dioxide produced during the combustion process and the inherent ash alkali, and (2) the lower temperatures at which the fluidized combustion process can be run.

The longer contact time allows for the reaction between coal ash alkali, oxygen, and SO_2 to be carried further towards completion. The lower temperatures allow for greater alkali utilization because the temperatures are below the dissociation temperature of the sulfates and below the ash fusion temperatures so that the alkali is not tied up in the glassy agglomerate. The sulfur retention percentages obtained in preliminary tests on one North Dakota lignite at the Grand Forks Energy Research Center range from 38 to 58% without ash reinjection; this section presents the results of the preliminary parameter study of this lignite from the Beulah mine together with emission data from six additional Western coals.

Experimental Apparatus and Procedure

Test Equipment: The fluidized-bed combustor (FBC) used for the study is shown in Figure 5.10.

Figure 5.10: Schematic of GFERC's 6-inch FBC

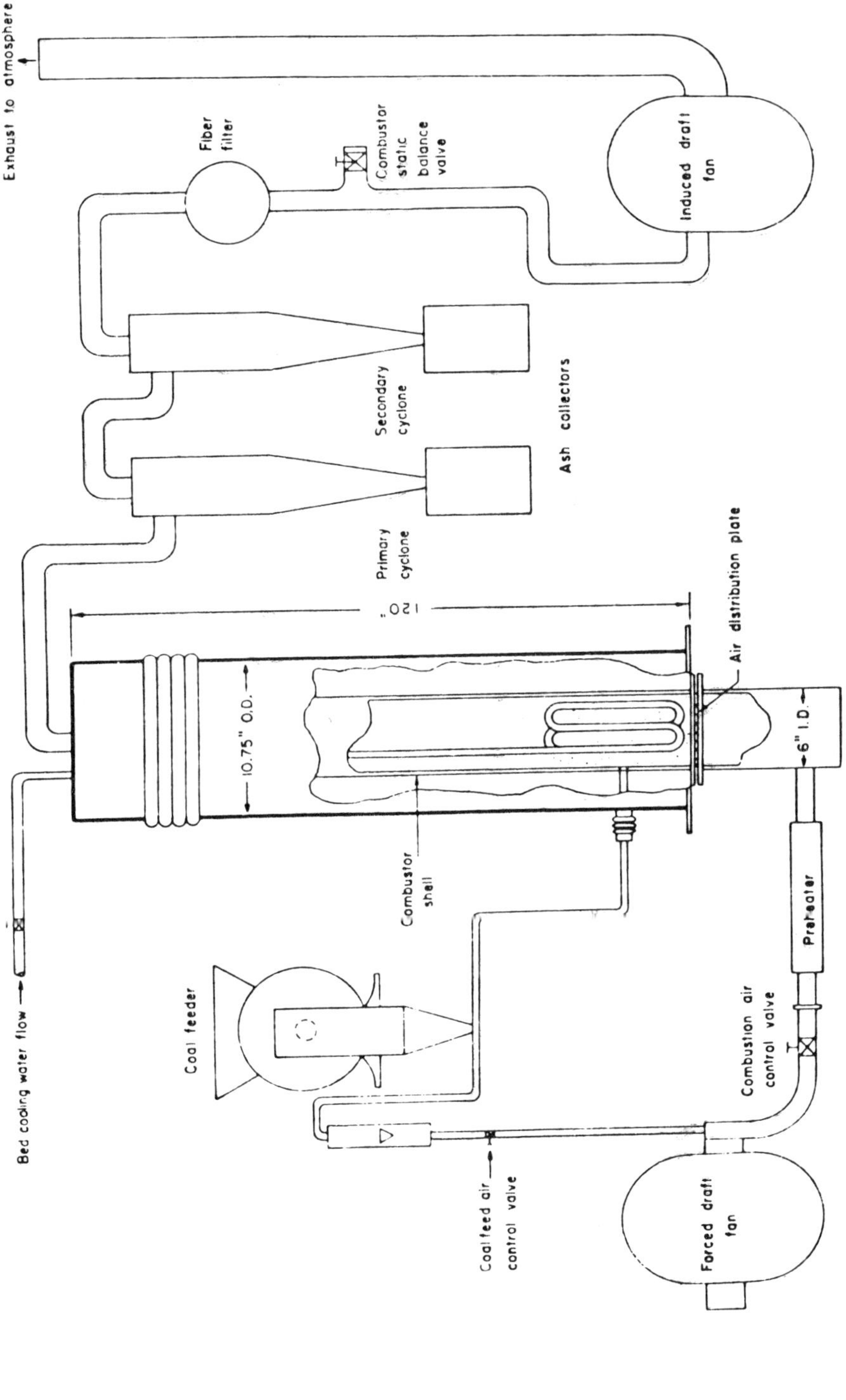

Source: NTIS GFERC/IC-77/1

The mechanical design of the unit is described in detail in a Master's Thesis (7) completed in December 1976. Results in the thesis study indicated very high sulfur retentions which could not be confirmed in later testing; the discrepancy was determined to relate to an error in sampling. The combustor shell is made from 6-inch, type 316, schedule 40 stainless steel pipe. It is 10 ft high, which allows sufficient freeboard above the bed to limit elutriation. The lower part of the combustor is wound with cooling coils on the outside, through which the water flow rate is manually controlled.

The combustor shell is enclosed in a stainless steel sheet metal skin which is filled with a loose insulation. Internal stainless steel cooling coils are the primary means for controlling temperature. The water flow rate to these coils is controlled by an automatic control valve which is operated from a thermocouple in the bed. Combustion air from a 7.5 hp positive displacement-type blower is electrically preheated both to provide ignition and to maintain a selected air preheat temperature during a combustion run.

The coal is metered to a pneumatic coal injector by a screw-type feeder. The rate of coal injection can be varied from about 10 to 30 lb/hr. The off-gases from the combustor are cooled by a series of heat exchangers to about 400°F. Gases are cleaned of particulate by a primary and secondary cyclone and a fiber filter before entering the induced draft fan, which is used to control the combustor static pressure. The gases are sampled once during each test for particulate loading, downstream from the combustor and ahead of the cyclone. The stack gases are monitored continuously for oxygen, carbon dioxide, carbon monoxide, nitric oxide, and sulfur dioxide. Oxygen concentrations are determined by an amperometric-type analyzer. Carbon monoxide and carbon dioxide concentrations are monitored by infrared analyzers. Nitric oxide emissions are analyzed by a chemiluminescent-type analyzer, and the sulfur dioxide concentrations are determined on a dry flue gas basis by an ultraviolet analyzer.

Procedure: The bed for the combustion tests on North Dakota lignite is a 99.5% pure, 30 mesh alundum grain (Al_2O_3). This material is added to the bed to a static height of about 8 inches, which becomes an expanded or fluidized-bed height of about 1.5 ft. The bed of alundum grain is fluidized with air which is passed through the electric air preheater until an average temperature of about 600°F is achieved in the bed.

When this temperature is reached, a small amount of coal is fed to the bed through the pneumatic injector. In this way, the bed is slowly brought up to 1000°F without an accumulation of coal in the bed. When the bed temperature reaches 1000°F, the coal feeder is turned on continuously, and a steady state condition is achieved. The air flow to the bed is adjusted to obtain the desired superficial gas velocity at the desired bed temperature, which is controlled by regulating the water flow to the heat transfer tubes immersed in the bed. The desired air-fuel ratio is achieved by adjusting the speed of the coal feeder screw to obtain the required flue gas oxygen concentration. Once steady state conditions are achieved, data are taken over about a 4-hour period.

Lignite Used for Experiments

The ultimate and proximate analysis of the Beulah coal used for these tests are shown in Tables 5.5 and 5.6, respectively.

Table 5.5: Beulah Lignite Ultimate Analysis

Components, %	As-Burned, %	Moisture-Free, %
Hydrogen	5.93	4.00
Carbon	47.20	64.77
Nitrogen	0.45	0.62
Sulfur	1.01	1.38
Oxygen	37.92	18.95
Ash	7.49	10.28
Heating value, Btu/lb	7,885	10.820

Source: NTIS GFERC/IC-77/1

Table 5.6: Beulah Lignite Proximate Analysis

Components, %	As-Burned, %	Moisture-Free, %
Moisture	27.13*	–
Volatile matter	30.84	42.46
Fixed carbon	34.44	47.26
Ash	7.49	10.28

*As-mined moisture was ~35%. Drying occurred during storage.

Source: NTIS GFERC/IC-77/1

The sulfur dioxide emission level for this lignite at 20% excess air, assuming no sulfur is captured by the ash alkali, would be about 1,172 ppm or 2.56 lb $SO_2/10^6$ Btu, therefore, this is a noncompliance coal. To meet the current Federal New Source Performance Standard of 1.2 lb $SO_2/10^6$ Btu about 54% of the coal sulfur must be retained. Table 5.7 shows the coal ash analysis for the Beulah lignite on a sulfur-free basis.

Table 5.7: Beulah Lignite Ash Analysis, SO_3-Free Basis

Silica, SiO_2	21.7
Aluminum oxide, Al_2O_3	14.4
Ferric oxide, Fe_2O_3	16.1
Titanium oxide, TiO_2	2.2
Phosphorus pentoxide, P_2O_5	0.8
Calcium oxide, CaO	29.9
Magnesium oxide, MgO	8.1
Sodium oxide, Na_2O	6.4
Potassium oxide, K_2O	0.4

Source: NTIS GFERC/IC-77/1

The calcium-to-sulfur ratio calculated from the above coal ash analysis and the coal ultimate analysis shown in Table 5.5 is 1.27. The total alkali (calcium, magnesium, and sodium) to sulfur ratio for this lignite is 1.99.

For the parameter study on bed temperature, air-fuel ratio, and superficial gas velocity, the lignite was crushed to minus 8 mesh. A sieve analysis of the coal is shown in Figure 5.11. Approximately 50% by weight of the particles are less than 800 microns (20 mesh). The fuel used for the tests on particle size was from the same batch of coal; sulfur content was essentially the same for all three sizes tested. The sizes tested were: the smallest size, minus 30 mesh which has an average size of about 300 microns; the middle size, which was used as the standard for the other parameter tests, minus 8 mesh equivalent to an average size of 800 microns; and the largest size, between 8 and 30 mesh, an average size of about 1,500 microns.

Figure 5.11: Beulah Minus 8 Mesh Particle Size Analysis

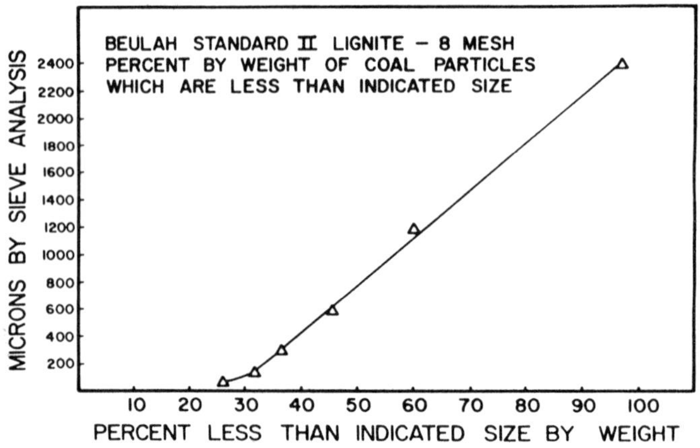

Source: NTIS GFERC/IC-77/1

Operational Variables

The average bed temperature was determined by the use of five thermocouples within the bed itself. The temperature range over which tests were conducted was from a low of 1200°F to a high of about 1500°F. One thermocouple in the bed which was reasonably representative of the average was selected to control the water flow to the internal cooling coils through an electropneumatic temperature controller.

The superficial gas velocity is calculated based on the flow of air, at the average bed temperature through an empty combustor. The velocities studied in this series of tests ranged from about 5 to 12 ft/sec at 1400°F. The cold air flow to the bottom of the bed is measured by an orifice flow meter ahead of the electric air preheater. During a run, the combustion air preheat is held at 400°F.

Air-fuel ratios varied from 9 to 60% excess air. The percentage of excess air is determined from the flue gas oxygen concentration, which is controlled by controlling the speed of the coal feeder while monitoring flue gas oxygen. The combustor pressure is maintained at about 6 inches wg below atmospheric for all the tests, and is controlled by regulating the amount of air bypassed to the induced draft fan.

Results of Parameter Study

Sulfur Retention: Figure 5.12 shows the percent of coal sulfur retained by the ash alkali when burning the Beulah lignite in the FBC over the range of average bed temperatures. All temperature tests were performed with approximately 30% excess air. The superficial gas velocity during all temperature parameter tests was about 7.3 ft/sec, and the coal feed rate was about 19 lb coal per hour. The maximum sulfur retention by the ash alkali occurred at an average bed temperature of about 1400°F, and was approximately 51%. The emission level at this condition was about 570 ppm, just slightly above the 1.2 lb $SO_2/10^6$ Btu Federal standard. The lowest sulfur retention, 38%, occurred at about 1500°F. The emission level in lb $SO_2/10^6$ Btu can be closely estimated for any point on this curve, for Beulah lignite, by the equation:

$$\text{(1)} \qquad \frac{\text{lb } SO_2}{10^6 \text{ Btu}} \cong 0.0256\,(100 - R)$$

where R is the sulfur retention in percent. The sulfur retention by the ash alkali of the Beulah lignite in pulverized coal combustion would be estimated to be about 30%. This sulfur retention is calculated from the empirical correlation (8) shown in the equation:

$$\text{(2)} \qquad R = 12.7\,\frac{CaO}{Al_2O_3} + 48.1\,\frac{Na_2O}{SiO_2} - 10.1$$

where CaO, Al_2O_3, Na_2O, and SiO_2 are the percentages of these components of the coal ash as shown in Table 5.7. In pulverized coal fired units, the alkali components calcium oxide and sodium oxide tend to retain the sulfur, while the aluminum oxide and silica make some of the alkali unreactive by reacting with the alkali.

Figure 5.12: Sulfur Retention Versus Average Bed Temperature

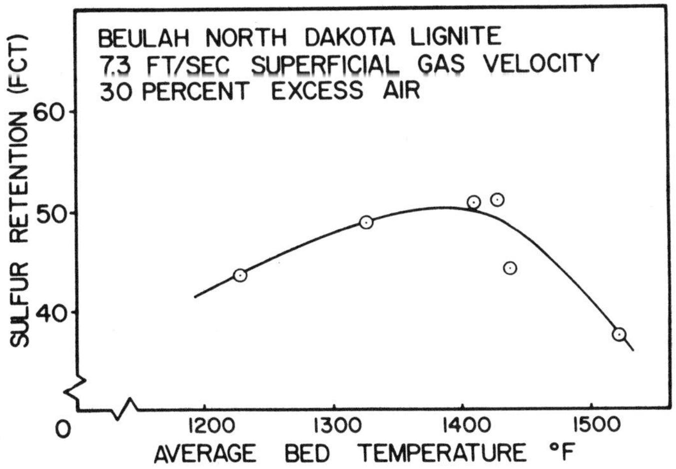

Source: NTIS GFERC/IC-77/1

The percentage of coal sulfur retained versus the air-fuel ratio is shown in Figure 5.13. The stoichiometric air-fuel ratio for the Beulah coal is, on a mass basis, 6.23 lb air/lb coal. This curve shows that the sulfur retention increases with the percentage of excess air. It is suspected that the higher concentration of oxygen in the bed allows the reaction of the calcium oxide, oxygen, and sulfur dioxide to go further toward completion. At an average bed temperature of 1400°F and approximately 61% excess air, about 58% of the coal sulfur was retained by the ash alkali. The sulfur dioxide emission at this condition was about 380 ppm, which is slightly less than 1.1 lb $SO_2/10^6$ Btu. Again, equation (1) can be used to estimate the emission level from the sulfur retention percentage indicated at any point in Figure 5.13.

Figure 5.13: Sulfur Retention Versus Air-Fuel Ratio

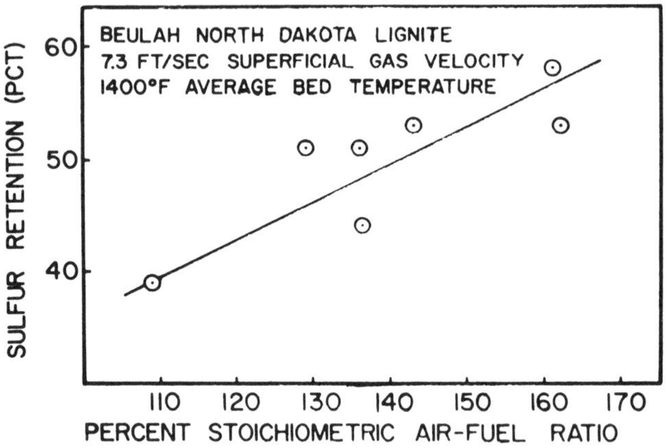

Source: NTIS GFERC/IC-77/1

The third variable tested in the parameter study tests was superficial gas velocity. This gas velocity is based on the average bed temperature during tests at approximately 1400°F. Data shown in Figure 5.14 indicates that gas velocities in the range studied have little effect on the retention of coal sulfur by the ash alkali. The slight decrease in retention with increasing gas velocity is most likely caused by the shorter retention time of the ash alkali in the reaction zone, because of the greater elutriation tendency. The tests in this series were all conducted with an average bed temperature of approximately 1400°F and an average air-fuel ratio of about 30% excess air. At about 7 ft/sec superficial gas velocity, the retention is 53%, and the emission level is 1.2 lb $SO_2/10^6$ Btu.

The last variable tested was coal particle size. Only three sizes of coal were tested. The first was the full range of sizes less than 8 mesh, the largest was between 8 and 30 mesh, and the smallest was everything less than 30 mesh. Figure 5.15 presents the sulfur retention data versus the average coal feed particle size. The change in retention over the range of particle sizes tested is very small and ranged from a low of about 53% to a high of 57%.

Figure 5.14: Sulfur Retention Versus Superficial Gas Velocity

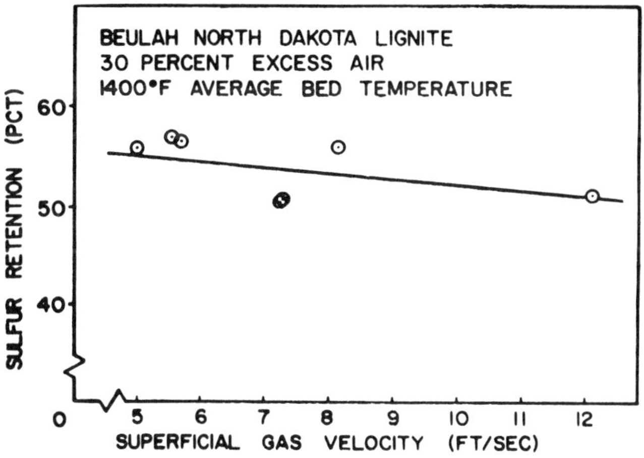

Source: NTIS GFERC/IC-77/1

Figure 5.15: Sulfur Retention Versus Coal Particle Size

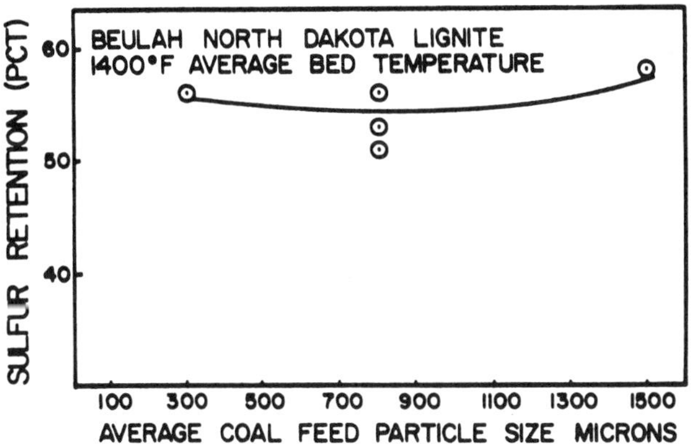

Source: NTIS GFERC/IC-77/1

Nitric Oxide Emission: If it is assumed that all or most of the nitric oxide emitted in fluidized-bed combustion comes from the fuel nitrogen and is not formed from the fixation of atmospheric nitrogen, the conversion of fuel nitrogen to nitric oxide versus the operational parameters can be plotted. The percent of fuel nitrogen as nitric oxide versus the average bed temperature is shown in Figure 5.16. The percent of fuel nitrogen converted to nitric oxide increases as the average bed temperature increases. This indicates that nitric oxide formation is dependent on temperature which is inconsistent with some other work done on nitric oxide production in fluidized-bed combustion. The emission level of nitric oxide can be determined at any point on this curve by equation:

$$\text{(3)} \qquad \text{lb } NO_x \text{ emitted} = 9.4 \times 10^{-3} N_e$$

where N_e is the percent of fuel nitrogen emitted. For example, at 1500°F, this curve indicates about 76% of the fuel nitrogen is emitted, which is equivalent to about 0.7 lb $NO_x/10^6$ Btu or equal to the New Source Performance Standards for bituminous coals.

Figure 5.16: Percent Fuel Nitrogen Emitted Versus Average Bed Temperature

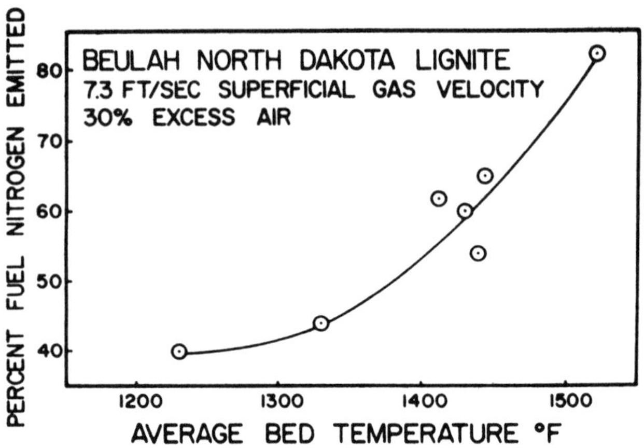

Source: NTIS GFERC/IC-77/1

The percent of fuel nitrogen emitted versus the percent of stoichiometric air-fuel ratio is presented in Figure 5.17. As has been demonstrated in other work both in FBC and pc-fired units, the amount of nitric oxide emitted increases as the available oxygen increases. The percentage of fuel nitrogen emitted as nitric oxide when burning the Beulah lignite in the fluidized-bed combustor at 1400°F and an average excess air level of 60% is about 65%. By the use of equation (3) for the Beulah lignite, this is equal to about 0.6 lb $NO_x/10^6$ Btu.

As shown in Figure 5.18, the superficial gas velocity through the bed has essentially no effect on the production of nitric oxide, or the percent of fuel nitrogen emitted as NO_x. The percent of fuel nitrogen emitted is constant at all gas velocities over the range tested at about 56% (0.5 lb $NO_x/10^6$ Btu).

Combustion Efficiency: Table 5.8 gives the analysis of the ash collected during a dust loading test while burning the Beulah coal. This analysis shows a 44.5% loss by weight during ignition of the ash at 800°C. This loss represents a total combustible carry-over with the fly ash of 3% by weight of the original feed coal. The combustion efficiency calculated for this particular run was 94%, which includes the combustible carry-over and the carbon monoxide in the flue gas. The data from all tests had not yet been analyzed, so combustion efficiencies for all the tests were not available.

Figure 5.17: Percent of Fuel Nitrogen Emitted Versus Air-Fuel Ratio

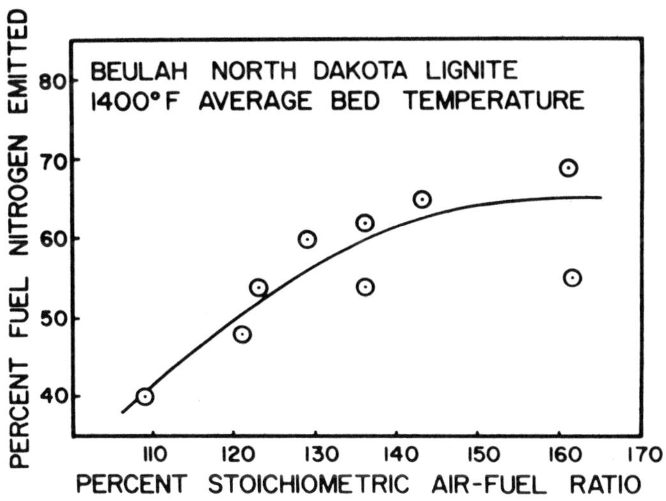

Source: NTIS GFERC/IC-77/1

Figure 5.18: Percent Fuel Nitrogen Emitted Versus Superficial Gas Velocity

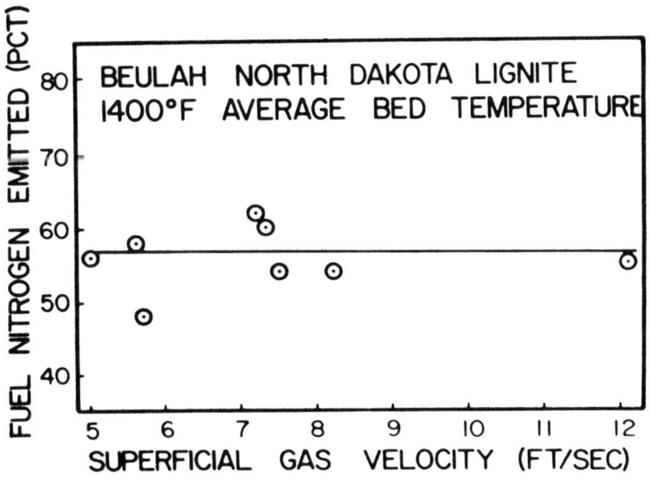

Source: NTIS GFERC/IC-77/1

Table 5.8: Typical Dust Loading

	Ash Analysis (%)	After Ignition and SO_3-Free
Loss on ignition	44.5	–
Silica	12.0	25.2
Aluminum oxide	7.7	16.2
Ferric oxide	8.5	17.8
Titanium oxide	0.2	0.5
Phosphorus pentoxide	0.4	0.8
Calcium oxide	12.8	26.7
Magnesium oxide	3.7	7.8
Sodium oxide	2.3	4.8
Potassium oxide	0.1	0.2
Sulfur trioxide	7.8	–

Source: NTIS GFERC/IC-77/1

Discussion

The tests conducted indicate that the percent of sulfur retention has a peak value at about 1400°F. This value differs from the peak sulfur retention temperature when burning Eastern coals with limestone additive reported to be about 1500°F. It is believed that the reason for the shift of this peak to a lower average temperature (1400°F) is that the high local temperatures around burning coal particles cause some of the alkali to be unreactive. At lower temperatures, the reaction kinetics of the formation of calcium sulfate probably limits sulfur retention, and sulfur retention could be increased by ash reinjection.

The increase in sulfur retention with increasing excess air could be caused by the increased availability of oxygen for the calcium, oxygen, sulfur dioxide reaction. The overall reaction is shown in the equation:

$$CaO + \tfrac{1}{2}O_2 + SO_2 \longrightarrow CaSO_4 \quad (4)$$

Tests with Eastern coals and limestone have shown no sulfur retention dependence on oxygen concentration for conditions where excess oxygen is present. That is, as long as oxidizing conditions are present, sulfur retention was not dependent on oxygen concentration.

The curve of sulfur retention versus superficial gas velocity as shown in Figure 5.14 shows little effect of superficial gas velocity on sulfur retention for the range of velocities tested. The slight decrease in retention with velocity is most likely caused by the greater elutriation of ash fines at the higher gas velocities.

The effect of no change in sulfur retention with coal particle size is not surprising because the sulfur content and ash composition of the various coal particle sizes are not very different, and the combustion of this range of coal particle sizes is expected to be completely diffusion controlled.

The increase in the amount of nitric oxide emitted with temperature means either that the mechanism of formation is temperature dependent, or that any mechanism that destroys NO_x, after its formation, is inversely dependent on temperature. The formation of nitric oxide from fuel nitrogen is not very temperature dependent [equation (5)] but the equilibrium reaction [equation (6)] which would tend to destroy NO at lower temperatures and high NO concentrations as seen in FBC, is temperature dependent (9).

$$(5) \qquad N_{fuel} + \tfrac{1}{2}O_2 \longrightarrow NO$$

$$(6) \qquad N_2 + O_2 \rightleftharpoons 2NO$$

Therefore, unless fixation of atmospheric nitrogen is taking place, it seems unlikely that the formation of NO would be greater at higher temperatures. At the lower bed temperatures, the production of carbon monoxide is greater than at high temperatures and the carry-over of carbon is also greater. Both conditions will reduce NO_x (9) after its formation causing a lower nitric oxide emission at low temperatures. It may be possible to gain insight into this situation by probing the bed for nitric oxide, and measuring NO_x emissions at different positions in the flue.

The increase in NO_x with increasing oxygen concentrations is most likely the result of the increased availability of oxygen for the reaction shown in equation (5), the oxidation of fuel nitrogen.

From Figure 5.18, it is seen that the emission of nitric oxide from the fluidized-bed combustion of lignite is not dependent on superficial gas velocity. This is not surprising because the formation of nitric oxide from fuel nitrogen is probably a very fast reaction which takes place as the coal is burned, therefore, retention time of the gas in the bed is expected to have little effect.

Combustion efficiencies of the order of 97 to 98% can be expected when ash is reinjected. The 94% combustion efficiency reflects the unburned carbon carry-over as well as the carbon monoxide concentration; with ash reinjection, it is expected that high carbon carry-over will not be a problem.

The retention of coal sulfur by the ash of coals and lignites appears to be most strongly related to the parameter of alkali-to-sulfur ratio as would be expected. The actual sulfur retentions for coals other than Beulah were not available at the time this paper was prepared. For some of the coals tested, like the Velva and Decker coals, the emission level is well below the 1.2 lb $SO_2/10^6$ Btu currently required by Federal standards.

Conclusions

(1) The alkali-to-sulfur ratio inherent in lignite and other Western coals is the most important variable affecting retention of coal sulfur.

(2) In burning Western lignite and subbituminous coals without limestone or ash reinjection, emissions for some coals exceed the Federal New Source Performance Standard by a small margin. With reinjection of ash to the fluidized-bed combustor, it should be possible to reduce the SO_2 emissions well below the current Federal standard of 1.2 lb $SO_2/10^6$ Btu.

(3) The temperature at which the fluid bed is operated can have a substantial effect on the emission of sulfur dioxide. It appears that a temperature of about 1400°F with lignite gives a maximum sulfur retention. With Eastern bituminous coals and limestone, the maximum retention occurs at a higher average temperature.
(4) Increasing excess air increases the retention of coal sulfur by the alkali of Western coal ash.
(5) Superficial gas velocity and average coal feed particle size have very small effects on the retention of coal sulfur by the ash alkali. At very high superficial gas velocities, the retention of sulfur will tend to decrease slightly because of the elutriation of fine ash particles.
(6) The increase in nitric oxide emission with increasing average bed temperature is probably caused by the reduction of NO at lower temperatures by CO and unburned coal, but at present, there are not enough data to be certain of the mechanism.
(7) The formation of NO from oxygen and nitrogen increases as the the air-fuel ratio increases and is likely caused by the increased availability of oxygen for the reaction.
(8) Nitric oxide production is not dependent on the superficial gas velocity in a fluidized-bed combustor burning lignite.

FLUIDIZED-BED COMBUSTION OF TEXAS LIGNITE AND LIGNITE REFUSE

Introduction

The Morgantown Energy Research Center of the U.S. Department of Energy began a program in June 1976 to examine the application of atmospheric fluidized-bed combustion (AFBC) technology to low-quality fuels. The purpose of this program is to evaluate the feasibility of burning low-quality fuels in AFBCs. Combustion characteristics of various low-quality fuels such as anthracite culm, subbituminous coals, and lignite coals have been examined in an 18-inch diameter fluidized-bed combustor. This section presents the results of recent tests using Texas lignite and lignite refuse. These tests included a total of 31 balance periods accumulating 409 hours of stabilized combustion with these lignite fuels.

Texas, which is the largest lignite-producing state, still generates almost 90% of its electricity from natural gas in spite of the state's rich lignite deposits. Identified coal resources in Texas have been assessed to be about 1% of the nation's total (10). It is estimated that more than half of the Texas coal resources are lignites occurring in three formations in the lower Tertiary of the Gulf Coastal Plain (Figure 2.4).

The Wilcox formation contains approximately 80% of the lignite resources, while smaller lignite deposits are found in the Yegua and Jackson formations. The best grade lignite occurs in the Wilcox formation north of the Colorado River, and the poorest grade occurs in the Jackson formation. Two samples of Texas lignite from the Wilcox formation were examined to assess this fuel as a potential feedstock for fluidized-bed combustors. One sample was from a commercially-mined deposit termed the Big Brown lignite. The second sample representing a younger lignite, Borehole 6, was obtained from large boreholes owned by Dow Chemical a few miles from the Big Brown commercial operation. Both of these samples came from an area near Fairfield, Texas, located in Freestone County.

Background

Experimental Facility: The Morgantown Energy Research Center fluidized-bed combustion tests were carried out in the No. 1 atmospheric pressure fluidized-bed combustor which is depicted schematically in Figure 5.19. It is basically a refractory-lined, cylindrical combustor having an internal diameter of 18 inches in the bed region which is 45 inches high and an expanded freeboard section of 24-inch diameter which is 27 inches high. The combustor is equipped with a horizontal, water cooled heat exchanger submerged in the bed which is comprised of six hairpin ½-inch schedule 160-type 316 stainless steel pipe. Each cooling loop is fed separately from a water manifold and any number from one to six of these tubes can be in service. Each tube has a surface area of 0.6 ft^2.

These tubes are used to obtain heat transfer coefficients in the bed and additionally to control bed temperatures. Because some of them may be out of service for extended periods of time, they must be physically able to withstand bed temperatures; alloy tubes are thus required. These in-bed tubes were not used for cooling in the lignite combustion tests. They were, however, present for fluid dynamic reasons.

The freeboard area is equipped with a single-pass heat exchanger with the tubes essentially horizontal in a zig-zag pattern around the periphery of the freeboard space. This heat exchanger is used to provide heat transfer coefficients and to control the exit gas temperature.

Fuel is metered from a hopper which is suspended from a weigh cell by a 1³⁄₁₆-inch variable speed screw. This fuel falls into a fuel injection chamber where it is entrained by compressed air at ambient temperature and transported through a 1-inch schedule 40 pipe through the base of the conical distributor to the nozzle. Velocity through the nozzle ranges from 60 to 80 fps.

Fluidizing air comes from a plant air supply system through a pressure regulating valve and panel operated control valve to the air plenum located beneath the conical distributor. From the plenum it passes through 95 ³⁄₁₆-inch holes arranged in seven equally spaced circles into the directional nozzles inside the air distributor. The nozzles are ⅛-inch stainless steel 90° elbows, each equipped with a drilled pipe plug on the discharge side. These elbows are welded onto the distributor plate and turned at various random angles all inclined toward the center of the cone. The pipe plugs are drilled to ⅛-inch on the first four circles radially outward from the cone apex and the plugs in the outer three circles are drilled to ¹⁄₁₆-inch. Open area amounts to 0.27% of the distributor plate area. At 4.9 fps fluid velocity, maximum pressure drop through the distributor and expanded bed amounted to 3 psig.

Flue gases exit the bed through a 5-inch stack and enter the No. 1 cyclone at approximately 500°F. The particulate removed in this cyclone is recycled back

Figure 5.19: Schematic of Fluidized-Bed Combustor

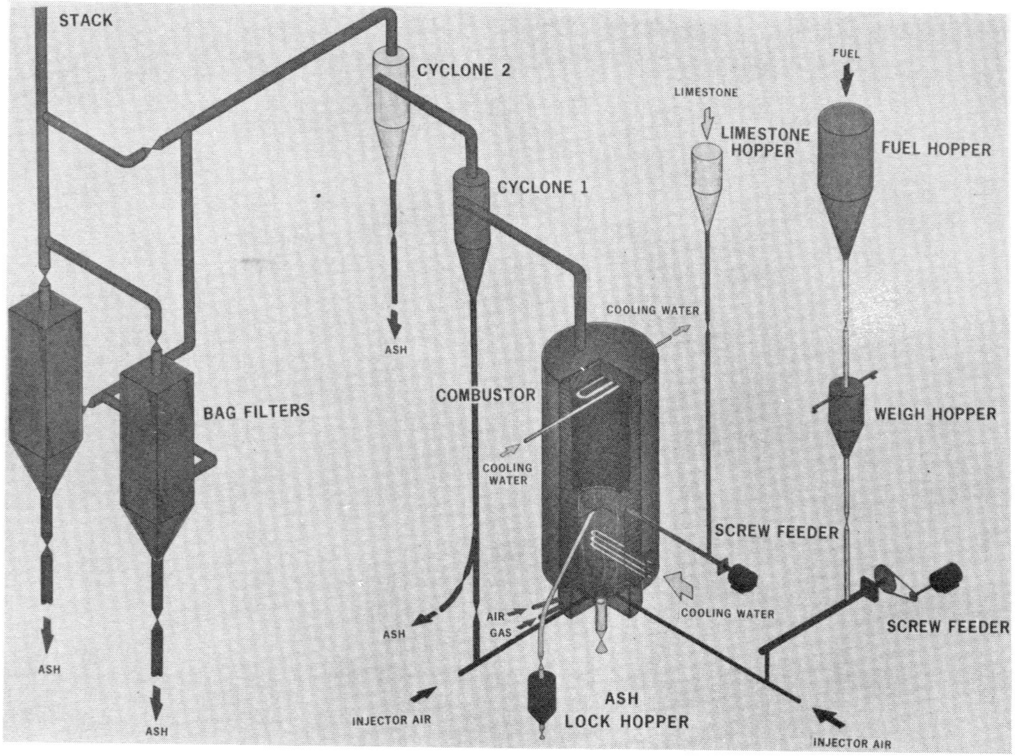

Source: NTIS MERC/RI-78/3

into the bed for additional carbon burnup. From the No. 1 cyclone the gas enters the No. 2 cyclone and the particulates removed here are weighed and analyzed.

From the No. 2 cyclone the gas exits into one of two parallel bag filters. Limestone (sorbent) when needed or used is fed via a 1³⁄₁₆-inch variable speed metering screw from a weigh hopper directly into the top of the expanded bed.

Ash was formerly extracted out of the combustor through a 3-inch schedule 40 pipe and a 3-inch feed screw located at the bottom of the conical distributor; however, excessive heating and warpage of the screw resulting from the high rate of bed removal required with the high ash content of the fuels, forced the installation of a 2-inch adjustable overflow tube which penetrates the side wall upward at about 30° angle. The opening in the overflow tube is normally set at the 36-inch level and this is used to control expanded bed height. The overflow tube discharges into a lock hopper which is periodically emptied. The overflow bed material is weighed and analyzed.

The stack gas is continuously sampled as it leaves the freeboard and passes through several filtering steps before being admitted to the analytical equipment. The gas monitoring (analytical) equipment includes the following instrumentation:

Compound	Principle of Operation	Instrument
CO	NDIR	MSA–LIRA Model 303
CO_2	NDIR	MSA–LIRA Model 1300
SO_2	Flame photometric	Meloy–FPD
NO-NO_x	Chemiluminescence	Thermo-Electron, series 10
O_2	Paramagnetic	Beckman
Total HC	Flame ionization	Beckman Model 400

Initially, long runs consuming five or six tons of fuel were made over a period of five 24-hour days. In later testing, the runs were broken down into 10- or 12-hour periods with 750 to 1,250 lb of fuel consumed per period. Velocities and/or bed temperatures are varied with each period.

Materials Tested: Fluidized-bed combustion experiments have been made using two Texas lignite coal samples, two additives, and various inert bed materials. Chemical analyses of the two samples, the Big Brown and Borehole 6, as well as the Borehole 6 refuse on an as-received basis are given in Table 5.9. Elemental analyses of the ash constituents are given in Table 5.10.

A portion of the Borehole 6 lignite was subjected to a water-washing, coal-cleaning process and separated into two sizes: lumps which were 1 x ¼-inch, and a fines fraction of minus ¼-inch x 0. The washery refuse had a size range of 1-inch x 0. Commercial utilization of lignites could well follow a similar preparation scheme in which the raw lignite would be cleaned to produce a 1% or less sulfur fuel for conventional utilization with higher sulfur refuse utilized in an AFBC for processing applications.

All fuels were crushed to minus ¼-inch before injection into the FBC. The Borehole 6 washed fines had the highest heating value, 9,400 Btu/lb, while the lignite refuse had the lowest heating value of 6,800 Btu/lb. Ash content ranged from a high of 31% in the refuse to a low of 8% in the washed lumps. Moisture content varied from approximately 11% in the Big Brown to 21% in the washed lumps. The sulfur content, as seen from Table 5.9, ranged from a low of 0.88% in the Borehole 6 washed lumps to a high of 2.34% in the Borehole 6 washery refuse.

Table 5.9: Analyses of Lignite Materials for Fluidized-Bed Combustion Tests As-Received

Constituent (%)	Materials Used: Texas Big Brown	Borehole #6	#6 Washed Lumps	#6 Washed Fines	#6 Washery Refuse
Total carbon	55.21	49.13	51.15	56.09	41.33
Hydrogen	5.00	4.66	5.83	5.80	4.14
Oxygen	14.98	13.88	11.69	11.53	9.75
Nitrogen	0.94	1.09	0.94	1.08	0.72
Sulfur	1.03	1.30	0.88	1.00	2.34
Moisture	10.65	16.42	21.26	15.26	11.13
Ash	12.19	13.52	8.25	9.24	30.59
Heating value, Btu/lb (as received)	9,170	8,237	8,920	9,375	6,805

Source: NTIS MERC/RI-78/3

Table 5.10: Elemental Analysis of Texas Lignite and Lignite Refuse Ash (HTA)

Fuel Type	Big Brown		Borehole #6: Washed Lumps	Washed Fines	Washery Refuse
Sample Source No.	CH 133	CH 455	CH 384	CH 395	CH 396
Constituent	(wt %)				
Silicon, SiO_2	40.08	44.83	32.40	39.82	34.90
Aluminum, Al_2O_3	16.58	12.40	15.11	16.06	10.71
Iron, Fe_2O_3	7.44	7.49	6.69	7.81	9.13
Calcium, CaO	14.65	13.10	18.42	13.39	24.31
Magnesium, MgO	4.54	2.11	5.50	2.12	4.25
Sodium, Na_2O	0.34	0.31	0.77	0.48	0.15
Potassium, K_2O	0.45	0.55	0.33	0.71	0.81
Phosphorus, P_2O_5	0.22	1.16	0.56	0.09	1.09
Titanium, TiO_2	1.46	1.02	1.44	1.26	0.61
Sulfur, SO_3	10.43	12.51	14.04	15.17	13.67

Source: NTIS MERC/RI-78/3

Both sorbents used in the tests, Greer limestone and Texas limestone, were screened to ⅛-inch x 0 before being fed to the top of the expanded bed. The chemical composition of Greer was 44.86% CaO, 38.37% CO_2. The Texas limestone composition was 50.90% CaO, 41.79% CO_2. High ash attrition and elutriation required the use of an inert bed material to maintain bed inventory. Inert material used included silicon dioxide, alumina, and ash material from the washery refuse, which contained a considerable amount of overburden. Inert material particle sizes were less than 12 mesh.

Operating Conditions: The purpose of the combustion studies was to investigate the feasibility of using lignite and lignite refuse in an AFBC and to assess the emission characteristics of this fuel. The combustion studies thus were carried out under the operating conditions such that a satisfactory and smooth operation could be sustained and the emission and operational characteristics which may be unique to this low-quality fuel could be determined. The range of operating conditions are listed in Table 5.11.

Table 5.11: Operating Conditions for Fluidized-Bed Combustion Tests of Texas Lignite

Average bed temperature, °F	1450-1750
Superficial gas velocity, ft/sec	3.0-6.0
Expanded bed depth, ft	1.5-3.0
Excess air, %	2.0-57.0
Fuel feed rate, lb/hr	37.0-108.0
Limestone feed rate, lb/hr	3.0-9.0

Source: NTIS MERC/RI-78/3

During the earlier studies, high average bed temperatures were also explored. It was found that the average bed temperature as high as 2000°F could be sustained without apparent clinkering.

Engineering data on carbon combustion, heat transfer, heat release and the emission of SO_2 and NO_x in the flue gas were measured under steady combustion conditions. These data were evaluated and analyzed. Multivariable effects and general data scatter became evident during the data analysis. The results of the tests are presented below.

Results of Combustion Tests

Carbon Combustion Efficiency: Carbon combustion efficiency is defined as one hundred minus the percentage of the combustible carbon lost with the fly ash. No carbon was lost through withdrawal of bed solids because bed inventory control did not require bed solids withdrawal during the tests. The variation in carbon combustion efficiency with excess oxygen in the flue gas and the average bed temperature are shown in Figures 5.20 and 5.21.

As seen, a 95% or better carbon combustion efficiency can be obtained over a fairly wide range of bed temperatures. Fluidizing velocities varied from 3 to 6 ft/sec. At excess air levels beyond 15% and at an average bed temperature above 1600°F, even large increases in excess air and raising temperature result in only slightly increasing carbon combustion efficiency. Excess air rate and bed temperature increase carbon combustion efficiency by decreasing the carry-over of particulate carbon and decreasing the carbon monoxide concentration in the flue gas.

Table 5.12 illustrates these effects at bed temperatures of 1470° to 1750°F and excess air rates of from 3.7 to 7.0%.

Figure 5.20: Carbon Combustion Efficiency Dependence on Excess Air

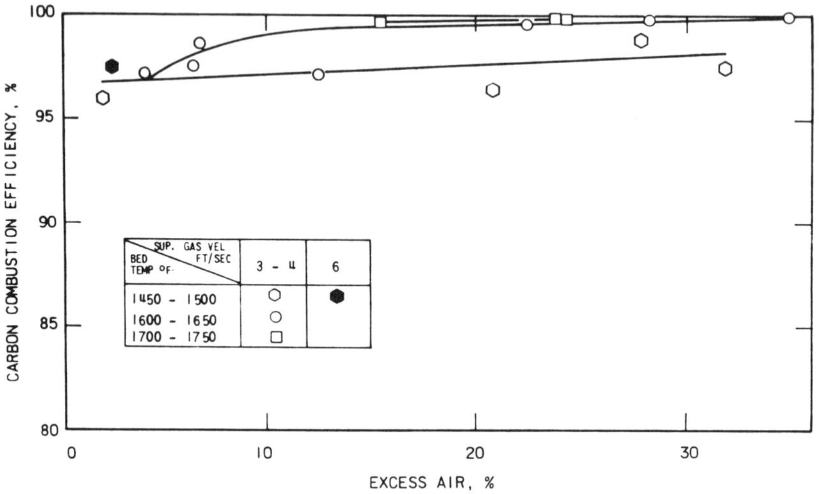

Source: NTIS MERC/RI-78/3

Figure 5.21: Carbon Combustion Efficiency Dependence on Bed Temperature

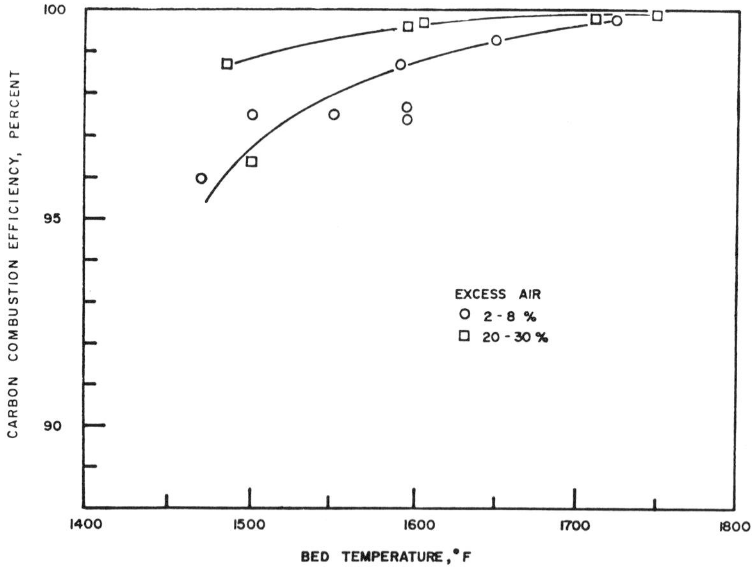

Source: NTIS MERC/RI-78/3

Table 5.12: Effects of Bed Temperature and Oxygen in Flue Gas on Carbon Monoxide and Particulate Carry-Over of Unburned Carbon

Run No.	Bed Temperature (°F)	Superficial Velocity (fps)	Oxygen in Flue Gas	Carbon in 2nd Cyclone Ash	Carbon in Bag Filter Ash	CO Flue Gas
			....	.... (%)	....	....
7/3	1600	3	7.00	1.6	2.7	0.06
8/7	1600	4	4.00	8.82	2.52	0.17
7/2	1470	3	4.50	16.20	10.81	0.40
8/8	1750	4	3.70	0.32	1.44	0.16

Source: NTIS MERC/RI-78/3

Sulfur Dioxide Emissions: The mineral constituents of Texas lignites and lignite refuse materials contain significant amounts of calcium which is effective in sulfur capture. This calcium reduces the amount of sorbent needed to maintain sulfur emissions within acceptable levels. The natural calcium contents of these lignites and lignite refuse materials were calculated to be equivalent to Ca/S mol ratio of 0.93 for Borehole 6; 0.99 for Borehole 6 washed lumps; 0.71 for Borehole 6 washed fines; and 1.86 for Borehole 6 washery refuse.

Dependence on Ca/S Ratio and Calcium Source — Sulfur dioxide emissions in the flue gas are plotted against the Ca/S mol ratio in Figure 5.22. The calcium content of the fuel is included in this ratio. The test data were obtained over a bed temperature range from 1450° to 1600°F and can be correlated fairly well by a single curve despite the variation in superficial gas velocities, bed depths, bed materials, sorbents, and the sulfur contents in the fuel feeds. The results indicate that current EPA sulfur dioxide emission limits can be met with relatively low sorbent addition rates and, in some cases, without sorbent addition. Calcium oxide in the lignite ash reduces and, in some cases, eliminates the need for sorbent addition. At bed temperature above 1600°F, no consistent effect of Ca/S mol ratio on sulfur dioxide emissions is evident.

The sulfur retention data are plotted as a function of the Ca/S mol ratio in Figure 5.23. As seen, a 90% sulfur retention can often be achieved at Ca/S mol ratio greater than 1.0 and at operating bed temperatures at or below 1600°F. Again, a single curve was used to correlate the test data. These combustion data also indicated that sulfur retention was better with Texas limestone than with Greer limestone over the bed temperature range of 1450° to 1600°F.

Effect of Bed Temperature — Without limestone addition, sulfur retention of Texas lignite and lignite refuse versus the bed temperature is shown in Figure 5.24. Although these data exhibit a great deal of scatter, a general trend can be observed showing that sulfur retention of Texas lignite and lignite refuse decreases with increasing bed temperatures over the temperature range of 1450° to 1750°F. A similar trend has also been reported by Goblirsch and Sondreal for North Dakota lignite (11).

Effect of Inert Bed Material — During the combustion tests with a pure quartz sand bed material, sulfur dioxide emissions increased in spite of the addition of sorbent. Observation of lignite ash by x-ray diffraction and SEM testing indicate an increase of calcium-aluminum-silicate parallel to the disappearance of the sulfur in the ash.

Figure 5.22: Sulfur Dioxide Emissions for Fluidized-Bed Combustion of Lignite and Lignite Refuse (Ca Derived from Ash and Limestone Addition)

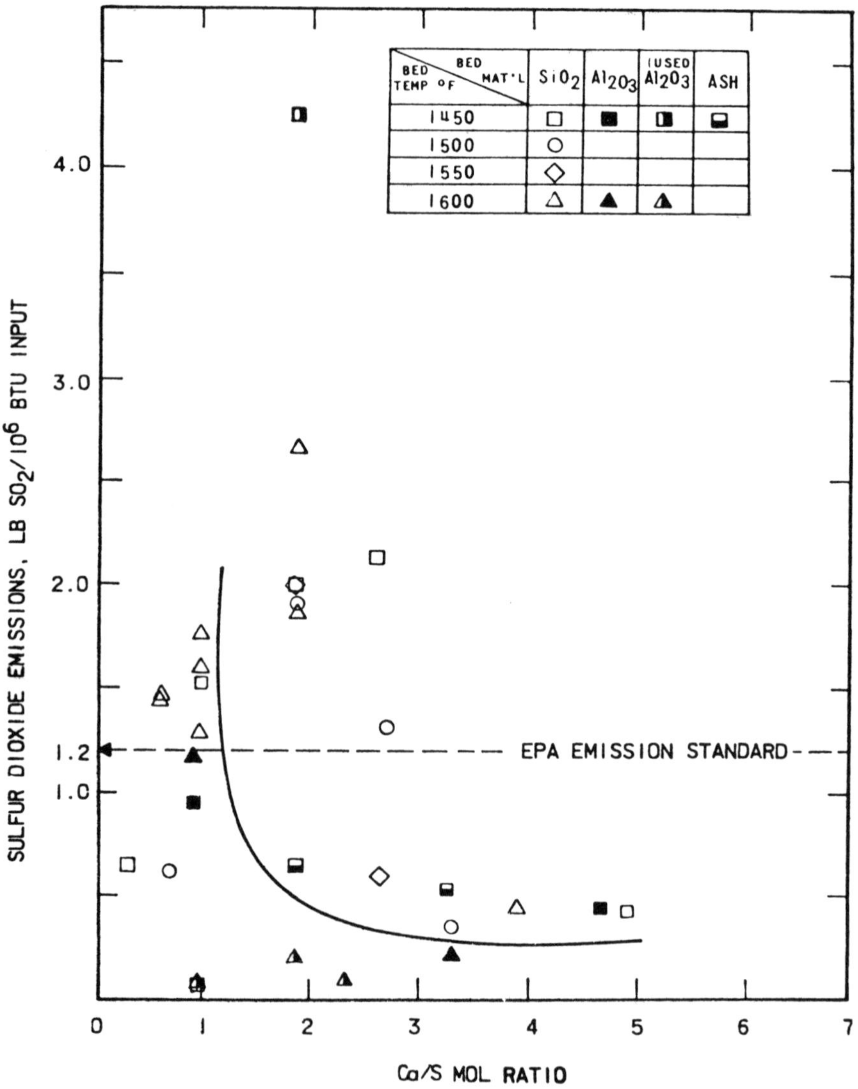

Source: NTIS MERC/RI-78/3

Figure 5.23: Sulfur Retention for Fluidized-Bed Combustion of Lignite and Lignite Refuse

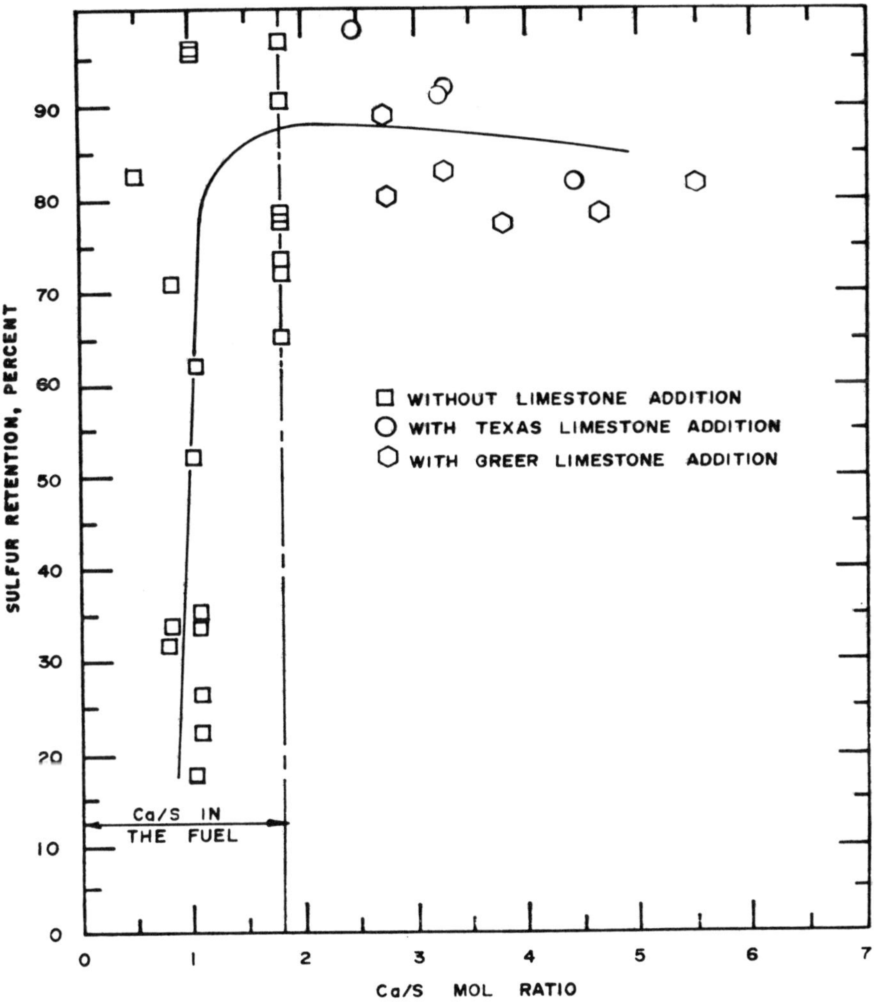

Source: NTIS MERC/RI-78/3

Figure 5.24: Sulfur Retention Decreases with Increasing Bed Temperature

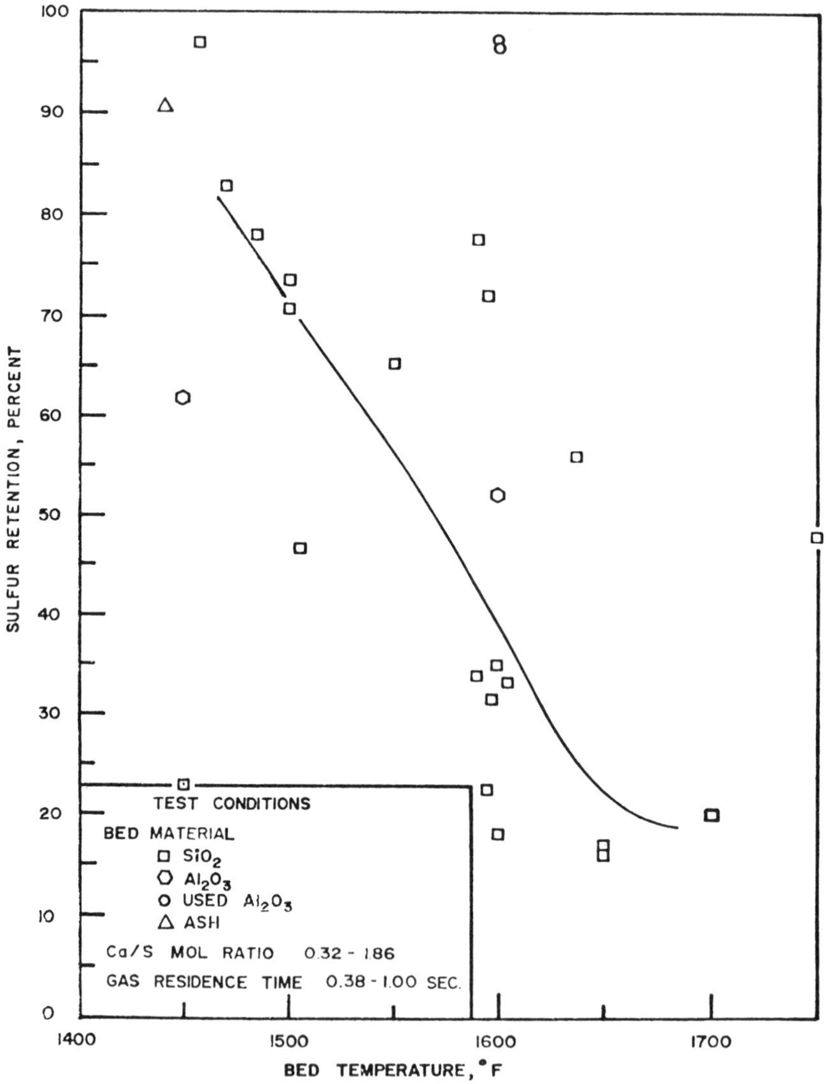

Source: NTIS MERC/RI-78/3

These preliminary results suggest that quartz sand bed material may affect the sulfur capture capability of a composite bed of lignite mineral, sorbent, and inert material. These results, however, are not conclusive.

Sulfur Retention on Lignite Ash – It has been previously reported (12) that sulfur is retained in the ash material while burning Western lignites. To confirm this hypothesis, a portion of the as-received lignite was tested for its sulfur retention ability. For this purpose, the coal was ashed by the standard ASTM method, but at temperatures which are realistic for fluidized-bed combustion. X-ray fluorescence analysis of the ashes confirmed that the ash material retains sulfur.

As shown in Figure 5.25, the SO_3 content of the ashes decreases from 15.45% at a 1600°F ashing temperature to 0.36% at 1850°F ashing temperature. Calcium in the ash participates in sulfur capture much like calcium in an added sorbent. The results confirm the FBC tests which indicated that there may be a temperature range where the lignite can be burned without added sorbents for sulfur emission control.

Nitrogen Oxide Emissions: Nitrogen oxide emission in the flue gas as measured during the combustion tests was low over the entire range of operating conditions. Values of NO_x concentration in the flue gas below EPA emission standards were generally observed depending on operating conditions as well as nitrogen content in the fuel. The fluidized-bed combustor was operated at a bed temperature between 1450° and 1750°F which is below the temperature level at which oxidation of atmospheric nitrogen occurs (13). The only significant form of NO_x in the flue gas was found to be nitric oxide. In Figure 5.26, nitric oxide emissions in terms of NO_2 are plotted versus oxygen in the flue gas. Effects of test variables on NO_x emissions are discussed below.

Effect of Fuel Nitrogen Content – Nitrogen content in the fuel also affects the nitric oxide emission. Table 5.13 indicates that NO emissions increase with increasing nitrogen content in the fuel.

Table 5.13: Effect of Nitrogen Content in Coal on Nitric Oxide Emissions

Run No.	Fuel	Nitrogen in Coal (%)	Bed Temperature (°F)	Superficial Velocity (fps)	Oxygen in Flue Gas (%)	Bed Material	Emission (ppm)
8/3	Borehole 6	1.01	1700	4	3.75	SiO_2	190
8/8	Washed lumps	0.94	1750	4	3.70	SiO_2	160
8/11	Refuse	0.72	1700	4	3.50	SiO_2	100

Source: NTIS MERC/RI-78/3

At a similar combustion condition, the Borehole 6, which contained 1.01% nitrogen, emitted almost twice as much NO as the Borehole 6 washery refuse.

Effect of Bed Temperature – The nitric oxide concentration measurements shown in Table 5.14 indicate that in the bed temperature of 1450° to 1650°F, NO emission is only slightly dependent upon bed temperature. Above 1650°F, nitric oxide decreases with increasing bed temperature. This is in contrast to the results obtained by Pope, Evans and Robbins (14) and Gibbs (15) which indicated that nitric oxide emission increases with bed temperature. The most likely explanation for the NO results in the present study is an interaction of NO and SO_2 in the bed. As mentioned earlier, sulfur retention capability of the composite fluidized-bed decreases as bed temperature increases above 1600°F. The nitric oxide concentration, in turn, likely decreases due to the effect of the sulfur dioxide concentration on the NO reduction reaction. This explanation, however, is not conclusive due to the limited data available and the errors involved in NO measurement.

Figure 5.25: Sulfur Retention in Lignite Ash Material at Fluidized-Bed Combustion Temperatures

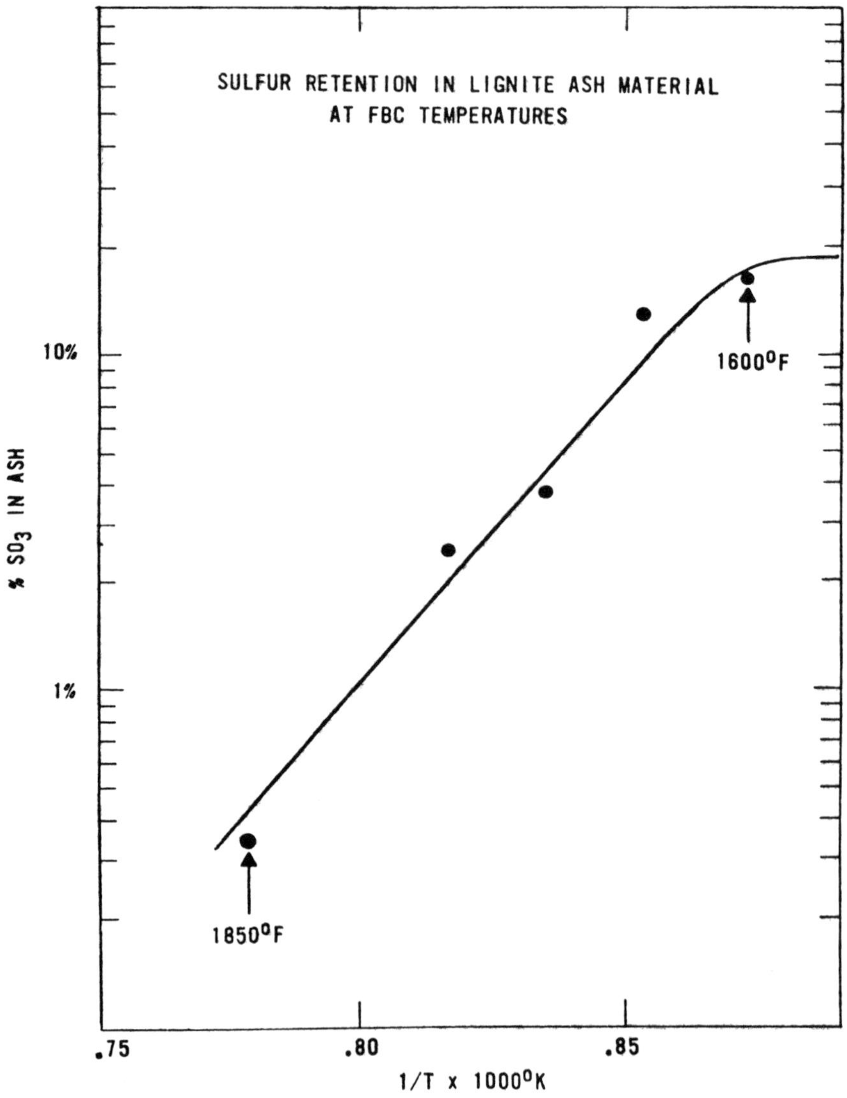

Source: NTIS MERC/RI-78/3

Figure 5.26: Effect of Oxygen in Flue Gas on Nitrogen Oxide Emissions

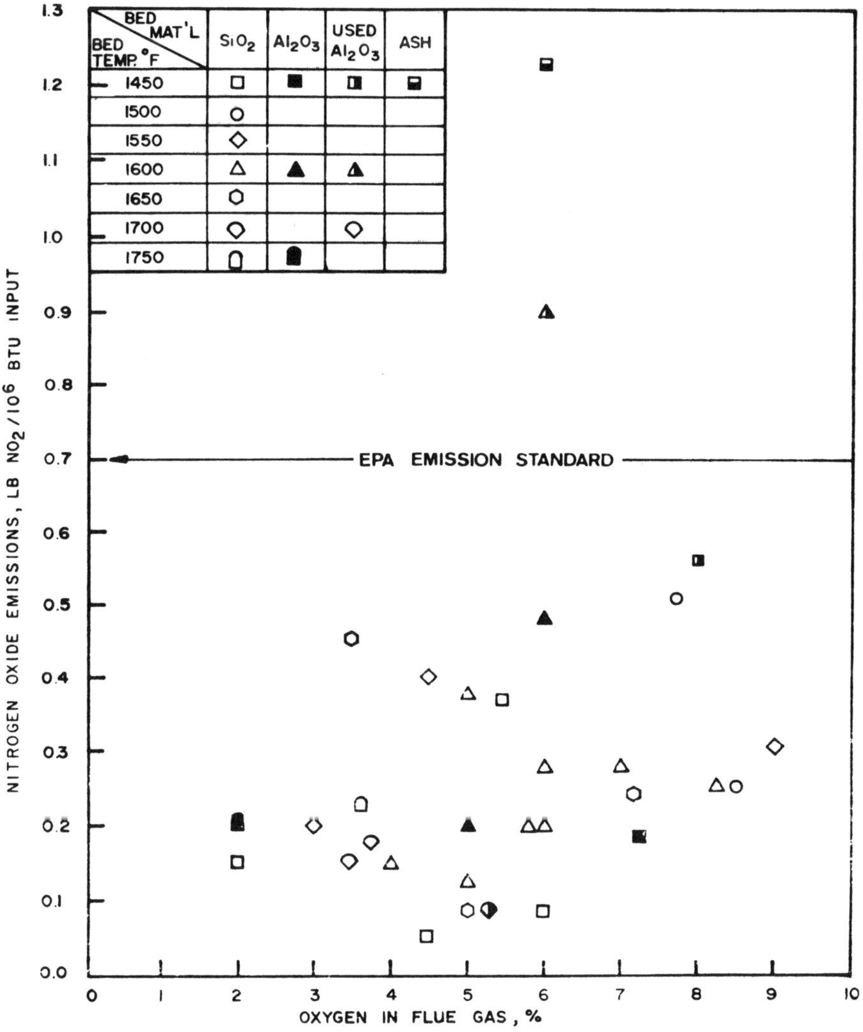

Source: NTIS MERC/RI-78/3

Table 5.14: Effect of Bed Temperature on Nitric Oxide Emissions

Run	Bed Temperature (° F)	O_2 in Flue Gas (%)	Superficial Velocity (ft/sec)	N_2 in Coal (%)	NO in Flue Gas (ppm)
8/1	1450	5.50	4	1.01	180
8/2	1600	5.00	4	1.01	210
8/9	1500	8.50	6	1.08	125
8/10	1650	7.10	6	1.08	150
8/3	1700	3.75	4	1.01	190
8/5	1550	3.00	4	1.01	210
8/11	1700	3.50	4	0.72	100
8/13	1650	3.75	4.5	0.72	240

Source: NTIS MERC/RI-78/3

Effect of Inert Bed Material – With SiO_2 and Al_2O_3 beds, the nitric oxide level in the flue gas increases with oxygen. The spread of these data suggests that bed material may affect the nitric oxide emission as illustrated in Table 5.15.

Table 5.15: Effects of Bed Material on Nitric Oxide Emissions

Run	Bed Material	O_2 in Flue Gas (%)	Bed Temperature (°F)	Superficial Velocity (ft/sec)	NO in Flue Gas (ppm)
8/1	SiO_2	5.50	1450	4	180
9/1	Al_2O_3	7.30	1450	4	85
9/6	used Al_2O_3	8.00	1450	4	405
9/8	ash	6.00	1450	3	600
8/2	SiO_2	5.00	1600	4	210
9/2	Al_2O_3	5.00	1600	4	104
9/5	used Al_2O_3	6.00	1600	4	440

Source: NTIS MERC/RI-78/3

It is apparent from these data that the used aluminum oxide and ash beds yielded substantially higher nitric oxide concentration levels in the flue gas. The nitric oxide emission increased from 210 ppm with SiO_2 as bed material to 440 ppm when the bed material was used aluminum oxide under similar operating conditions. A consistently lower NO emission from fresh Al_2O_3 beds was also observed throughout the combustion tests. The reduction of NO concentration in the flue gas may be due to a catalytic effect of SiO_2 and Al_2O_3.

However, experiments carried out by Argonne (16) with the injection of Al_2O_3 into the fluidized bed at a rate of several pounds per hour indicated that NO concentration in the flue gas was not affected by the addition of Al_2O_3. No conclusion on the effect of bed material on NO emissions has been suggested for the complicated results.

Effect of Superficial Velocity – The NO emission decreases slightly with increasing superficial velocity. Table 5.16 illustrates this slight dependence. At a given bed temperature of 1500°F and oxygen content of approximately 9% in the flue gas, an increase of superficial velocity from 3 to 6 ft/sec decreased the nitric oxide emission from 150 to 125 ppm. This effect of superficial gas velocity on nitric oxide emissions is consistent with data obtained by Argonne (16).

Table 5.16: Effect of Superficial Velocity on Nitric Oxide Emissions

Run	Bed Material	O_2 in Flue Gas (%)	Bed Temperature (°F)	Superficial Velocity (ft/sec)	NO in Flue Gas (ppm)
7/5	SiO_2	9.10	1500	3	150
8/9	SiO_2	8.50	1500	6	125
7/3	SiO_2	7.00	1600	3	168
8/10	SiO_2	7.10	1650	6	150

Source: NTIS MERC/RI-78/3

Temperature Distribution in the Fluidized-Bed Combustor: One of the problems encountered in fluidized-bed combustor operation is ash clinkering which is often caused by localized high temperature zones. Maintaining uniform temperature distributions within a bed will usually prevent ash agglomeration. During the fluidized-bed combustion tests on Texas lignite and lignite refuse, temperatures were measured at 10, 23, 28, 56 and 72 inches above the apex of the conical distributor plate. Results for runs 7/2, 7/3, 8/7 and 8/8 are listed in Table 5.17.

Table 5.17: Temperature Distribution in the Fluidized-Bed Combustor

	. . . . In Bed			. . . . Above Bed		
Run No	Thermo-couple 21 19" Above Plate (°F)	Thermo-couple 1 23" Above Plate (°F)	Thermo-couple 2 28.5" Above Plate (°F)	Thermo-couple 7 56" Above Plate (°F)	Thermo-couple 11 Top of Combustor (°F)	Bed Depth (in)
7/2	1470	1476	1477	1420	1090	30
7/3	1587	1594	1586	1401	1025	30
8/7	1585	1588	1588	1467	1253	36
8/8	1741	1748	1745	1521	1288	36

Source: NTIS MERC/RI-78/3

Comparison of the in-bed thermocouple numbers 21, 1 and 2 shows a maximum bed temperature variation of about 8°F along the vertical direction when the bed was fully fluidized. This small temperature gradient along the vertical direction indicates very rapid solid mixing in the bed.

Freeboard combustion is undesirable because it tends to increase emissions and leads to excessive temperature in the freeboard. As a consequence, ash could be softened and possibly foul downstream surfaces.

Freeboard temperature measurements near the bed surface differed only slightly from the average bed temperature whichiimplied that there was very little or no combustion taking place near the surface and in the lower freeboard.

Heat Transfer Coefficient and Heat Release Data: During the combustion test of Texas lignite and lignite refuse, in-bed heat exchanger tubes could not be used due to excess heat loss from the relatively small pilot plant. Temperature control with in-bed heat exchanger tubes in operation was judged to be too difficult. Freeboard heat exchangers were, however, employed throughout the tests. Average overall freeboard heat transfer coefficients were found to be approximately 10 Btu/ft^2-hr-°F.

The volumetric heat release rate in the tests ranged from a low of 0.08 x 10^6 Btu/ft^3-hr for the refuse at an expanded bed depth of 36 inches to a high of 0.20 x 10^6 Btu/ft^3-hr for the Borehole 6 at an expanded bed depth of 18 inches. Comparison of these data with the heat release rate obtained in burning high quality coal in a fluidized-bed combustor shows that the volumetric heat release rate for lignite can be as good as higher grade coal (17).

Particle Size Distribution: Lignite tends to crumble and disintegrate very easily upon losing its moisture. Knowledge of this unique characteristic is essential to the success in particulate emission control and to achieve satisfactory combustion efficiency. For the design of lignite coal FBC, data on particle size distribution of the entrained solids from the fluidized-bed is very important. Figure 5.27 shows typical distributions for the solid materials in the input and output streams during the test of lignite Borehole 6 at a superficial gas velocity of 4 ft/sec. This particular run was made without limestone addition. Average particle sizes for the streams in terms of weight means were 1,350 μ for the fuel feed, 820 μ for the spent bed material, 41 μ for the primary cyclone ash, 20 μ for the secondary cyclone ash and 3.5 μ for the bag filter ash.

Energy Balance: Energy balance analyses for the lignite and lignite refuse tests show that rather large amounts of heat (22 to 46% of the total heat input) are lost during the fluidized-bed combustion process. This excessive heat loss is due to the small size of the combustor and its large surface area to volume ratio coupled with the high temperature and high combustion intensity used in the tests. In a large combustor, the useful heat available in the bed zone, hence, would approach the sum of the radiation losses and that extracted from the submerged bed heat exchanger. It is evident from the energy balance analyses that a large fraction of heat is lost in the flue gas. Some provision for recovery of the heat in the effluent stream would undoubtedly benefit the overall efficiency.

In addition to the heat losses through radiation and flue gas, approximately 10% of the fuel heat input is spent in coal moisture vaporization which results from a relatively high moisture content in the lignite coal. In the design of lignite FBC, the heat leaving the combustor in the flue gas could be used for drying the lignite or perhaps preheating the air.

Analysis of Ash Content of Spent Bed Material and Entrained Solids: The spent bed material removed from the combustor contained less than 1% of combustible carbon; however, carbon content in the entrained solids varied from less than 1 to 9%. In general, the fly ash collected in the second stage cyclone contained a higher carbon concentration than the fine particulate collected from the bag filters. The difference in combustible carbon content per second stage cyclone fly ash and the fine particulate from bag filters can be attributed to particle size. The finer particles (see Figure 5.27 for particle size distributions) have a higher specific surface area and, therefore, reacted more quickly and more thoroughly than the larger particles retained by the second stage cyclone. Comparison of the mineral content of bed ash material with anthracite refuse (18) indicates that a fairly low concentration of sodium and potassium is present in the bed ash of Texas lignite and lignite refuse. This rather low concentration of sodium and potassium in the lignite spent bed material perhaps accounted for the fact that a higher bed temperature could be sustained without clinker problems.

Figure 5.27: Typical Particle Size Distribution for Lignite Combustion

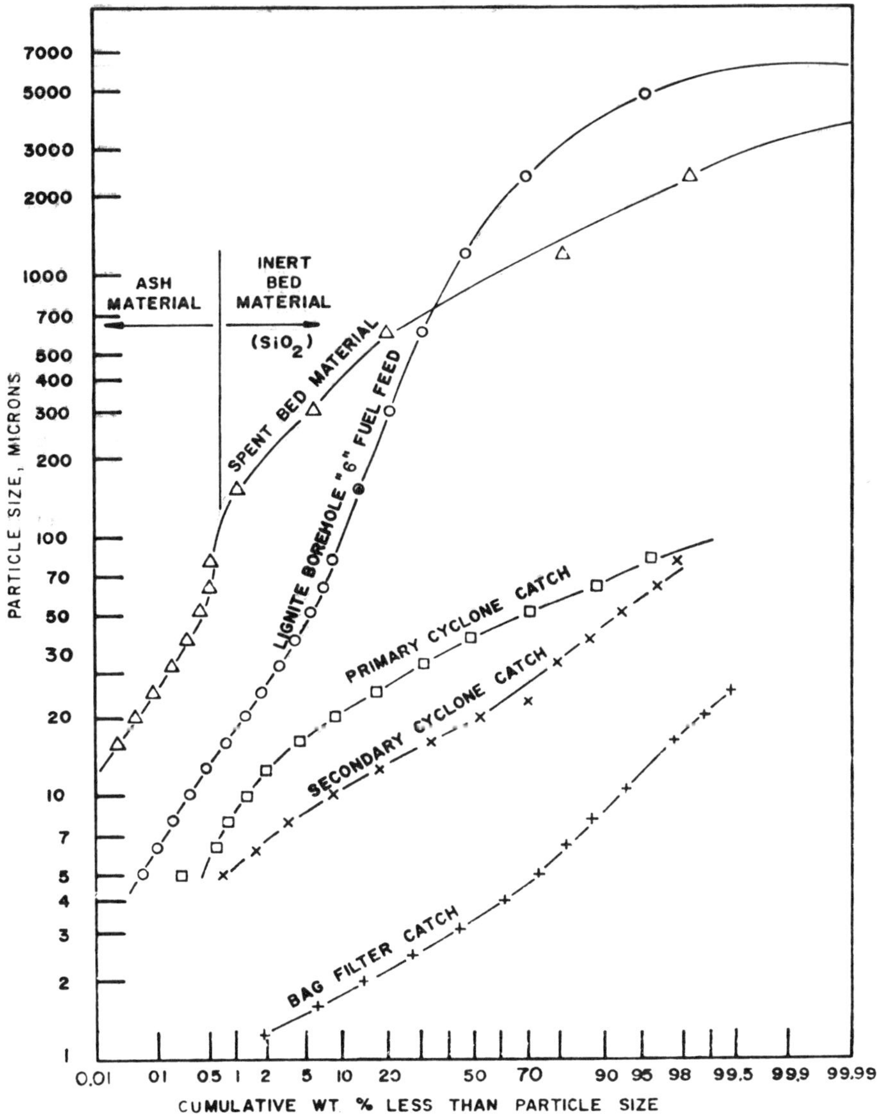

Source: NTIS MERC/RI-78/3

Conclusions

Two samples of Texas lignite, the Big Brown and the Borehole 6, both from the Wilcox formation have been successfully burned in the 18-inch diameter fluidized-bed combustor under a wide range of combustion conditions. The only significant problem encountered throughout the combustion tests was a relatively large amount of very fine ash particles produced and elutriated due to the high attrition rate of this fuel which, however, was solved readily by using an attrition resistant bed material. This low-grade coal, once considered little more than a high-class dirt, can be utilized as a feed stock for fluidized-bed combustors.

Based on the test data, several conclusions can be drawn from the present feasibility studies on combustion of Texas lignite and lignite refuse:

(1) With the recycling of primary cyclone ash, 95% or better carbon combustion efficiency is attainable.
(2) Current EPA sulfur dioxide emission standards can be met with little and, in some cases, no sorbent addition provided that the fluidized-bed combustor operates at bed temperatures at or below 1600°F.
(3) The data in this study indicate that a 90% sulfur retention may be obtainable with Ca/S mol ratios equal to or greater than 1.0 with bed temperatures at or below 1600°F.
(4) Nitrogen oxide concentration in the flue gas is low and is dependent upon the excess oxygen.
(5) The high volumetric heat release rates compare favorably with rates obtained in burning higher quality coal.
(6) Based on the combustion test data, the atmospheric fluidized-bed combustion technique offers a promising way to utilize Texas lignite and lignite refuse.

EXPERIENCE IN THE COMBUSTION OF BROWN COALS FROM THE LATROBE VALLEY (AUSTRALIA)

Introduction

The State Electricity Commission of Victoria (SECV), when constituted in 1918, was given responsibility for the development of the brown coal resources of Victoria for electricity and solid fuel production. Continuous large scale development of the brown coal fields of the Latrobe Valley has provided the basis for the power and fuel undertakings of the SECV that comprise three opencuts, one briquet factory, three operating power stations, and a further power station in which the first unit has been commissioned. Yallourn Power Station has a mixture of range connected plant and units of 120-MW capacity and has a total capability of 546 MW. Hazelwood consists of eight 200-MW units, while the 350-MW unit recently commissioned at Yallourn W Power Station is the first of four 350-MW installations. Total SECV investment in the Latrobe Valley exceeds $650 million.

Over the history of the SECV to June 1972, a total of 425 million tons of brown coal has been mined, and in 1971–72, production was 22 million tons for the year. Generation capacity was expected to increase to over 3,000 MW in

1974 and coal production to rise progressively to about 33 million tons per year. Total coal reserves of 1,100 million tons will have been committed for the remaining life of these plants.

An extensive program of geological investigations carried out over more than 25 years by the SECV has defined the extent of the Latrobe Valley brown coal deposits and the principal areas most suitable for opencut winning.

Figure 5.28 shows a plan of the major structures and of the concealed outcrops of the brown coal seams that have been identified in the different areas. On a geological basis, the coal reserves of the Latrobe Valley have been calculated as 47,500 million tons proved and 37,500 million tons inferred.

Figure 5.28: Brown Coal Deposits of the Latrobe Valley

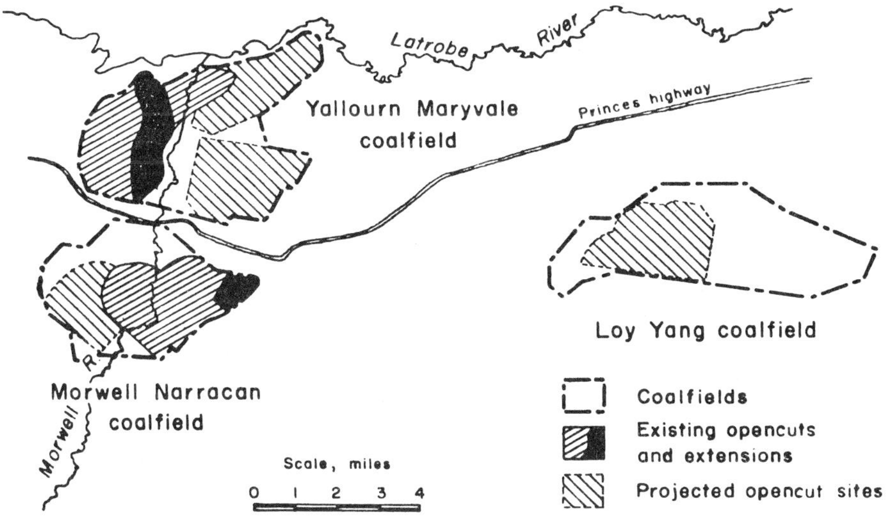

Source: IC 8650

Of the proved resources of coal, some 29,000 million tons occur with less than 100 ft of overburden over the uppermost seam. However, when provision is made for factors such as present townships, rivers, realistic opencut planning, and a limitation on cost of coal winning to approximately present cost, these resources are reduced to about 10,000 million tons.

Two major difficulties have been encountered in the use of brown coals from the Yallourn Opencut and the Morwell Opencut in power station boilers. The first boilers installed in 1924 to generate power from Yallourn brown coal were grate fired, and some difficulties were experienced in burning the coal, which has a normal moisture content of 67% but can be as wet as 73%. The problem of unstable combustion was encountered in serious proportions when the mill-fired Yallourn C boilers (Figure 5.29), producing 200,000 lb of steam per hour, were commissioned in 1954; despite a number of significant developments in handling and firing, flame-outs can still occur in the 120-MW Yallourn E boilers.

Figure 5.29: Comparative Sizes of Latrobe Valley Boiler Plants Fired with Pulverized Coal

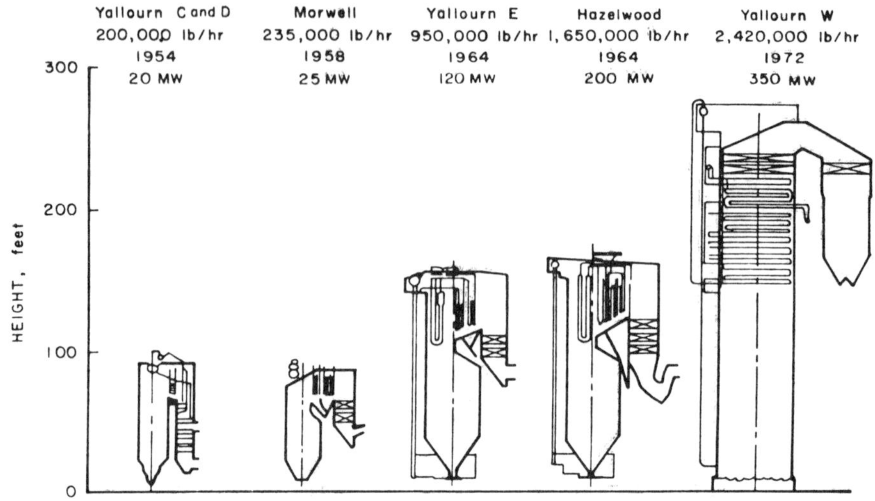

Source: IC 8650

Since the application of pulverized fuel firing, difficulties have been experienced from ash depositing on heating surfaces; at Yallourn, the boilers were equipped extensively for on-load cleaning with steam soot blowing in the furnace, superheater, and economizer regions. Shot cleaning was used in the economizers of the smaller boilers. Equipment and practices were developed to enable the boilers to operate between annual overhauls without an intermediate off-load cleaning outage, but this availability was achieved at the expense of boiler efficiency.

In 1958, very severe fouling was encountered with the first use of coal from the Morwell Opencut. Boiler deposits from Morwell coal formed more rapidly and (particularly in furnaces and superheaters) were much more difficult to remove than those from Yallourn coal. The first 200-MW boiler at Hazelwood Power Station could be operated initially for less than 500 hours before cleaning although 8,000 hours without recourse to off-load cleaning had been specified. A development that has contributed significantly to the alleviation of this problem has been the application of water soot blowing to the furnace walls of the boilers.

Boiler Operation

Combustion Stability: The problem of unstable combustion in mill-fired boilers was first encountered when the Yallourn C boilers were commissioned. The fuel used has a normal moisture content of about 67%, which is considerably higher than that of the German coals for which the firing equipment had originally been designed. Steam raising, running the turbine up to speed, and synchronizing were achieved using auxiliary burner equipment that operated on pulverized coal with approximately 15% moisture. The main mills were then brought into service, but ignition was lost when the auxiliary firing was shut down with the furnace alternately blacking out completely and then reigniting with heavy pressure surges. To keep the plant in commercial operation, it was necessary to run auxiliary firing equipment continuously to maintain ignition on the main burners. Even when operating in this manner, excessive furnace pulsations were still experienced.

Development of Combustion Systems: Different burner designs were tried as successive boilers were commissioned, and the first modification to improve conditions was the installation of large areas of refractory around the burners themselves. Refractory "belts" were subsequently fitted at burner level around the furnace of the four boilers then in service, and for the first time no auxiliary firing was necessary, although furnace pulsations were still excessive.

Finally, cyclones were fitted in the fuel ducts to enrich the fuel/gas mixture delivered to the lower part of each burner. With an enriched mixture containing about 80% of the fuel and one-third of the conveying gas introduced through the lower half of each burner and the remainder admitted through the upper half, a strong stable flame was obtained at the lower burner and satisfactory operating conditions could be maintained. All boilers in C station and in D station were equipped with these separators, and it was found possible to remove the refractory previously installed.

Investigations leading to the solution of the combustion instability problem for Yallourn C and D boilers have been described by Roberts (19) and Kennedy and Warren (20). This fuel separation system, pioneered at Yallourn, has since been adopted in plants overseas as an effective technique when dealing with brown coals that are difficult to burn.

The 120-MW boilers of the Yallourn E Station were equipped with exhausters capable of varying the gas/fuel ratio of the mixture entering the furnace; cyclones for further enriching the mixture were not provided. However, when burning wetter than average coal, fairly wide fluctuations in furnace pressure were experienced with these boilers, and auxiliary firing was used fairly frequently to stabilize combustion. Angled plates fitted in the fuel ducts above the exhausters resulted in an improvement by changing the distribution of fuel between the two openings of each burner group; "concentrators" of this type were eventually fitted to all ducts on both boilers. A test carried out with wetted coal indicated that combustion stability should be satisfactory with coals of moisture contents up to approximately 71%.

However, further difficulties were experienced with fluctuating furnace pressures, and loss of ignition occurred on a number of occasions. During one period of operation, it was found necessary to avoid coal from the base of the Yallourn seam that could have relatively high ash contents; ultimately it was decided to install cyclone separators similar to those developed for C and D stations. This modification resulted in significant improvements in boiler operation, but it has not completely eliminated periods of instability; some flame-outs in E station have occurred that cannot be related to changes in moisture or ash contents of the coal. In contrast, combustion instability has not occurred in power stations supplied with the 63%-moisture Morwell coal.

Coal Property Investigations: Because of the continuing trouble experienced with Yallourn E and the prospect of the 350-MW Yallourn W boilers operating on the same Yallourn coal, intensive investigations are being conducted on the coal itself to identify the coal properties that contribute to combustion instability. These studies have ranged from the ignition of single particles of brown coal to tests in a 75-lb/hr experimental furnace.

A study of the ignition of single particles of brown coal in a hot gas stream has established, both from a mathematical model and experimentally, that ignition is affected by such physical factors as particle size, gas temperature, oxygen partial pressure, and moisture content (21)(22). However, with these physical factors controlled at set levels, different Latrobe Valley brown coals still display wide variations in ignition behavior. Morwell coals are found to ignite more readily than Yallourn coals, and some areas of coal in the Yallourn Opencut with particularly poor ignition characteristics have been detected. Samples from bores in an undeveloped field (Loy Yang) were tested for comparison and found to be generally less reactive than the Morwell sample but more reactive than Yallourn samples.

These findings have been duplicated with a second bench-scale test that involves timing the ignition of a 0.5-gram sample of milled coal placed in a muffle furnace at 500°C. This simple test has been used for extensive surveys of the variation in ignition characteristics in the Yallourn Opencut.

Investigations in the 75-lb/hr pulverized-fuel experimental furnace at the Herman Central Scientific Laboratory (23) of the SECV have shown that furnace pressure fluctuation and effective flame radiation temperature can be used as measures of flame stability, and correlations obtained with water addition to the flame are consistent with the data obtained in early trials on Yallourn C Station boilers by Roberts (Figure 5.30). The stability of the flame in the experimental furnace to water addition was dependent on coal source (Figure 5.31) and also decreased with increasing fuel particle size; Morwell flames were more stable than those of Yallourn with Loy Yang intermediate between the two. These observed relative flame stabilities support the findings of the bench-scale ignition tests and are generally consistent with the operating characteristics of commercial boilers burning Yallourn and Morwell coals.

It has been shown (24)(25) that coal source is a factor that influences the particle size and moisture content of the pulverized fuel resulting from the milling and drying process. Thus, the milling and drying characteristics of coal can also influence combustion stability.

Ignition and flame stability investigations have been extended in an attempt to explain the poor ignition phenomenon and identify areas of poorly igniting coal within the Yallourn Opencut. It has been found that the ignition characteristics are influenced by the inorganic constituents and the petrology of the coal in a particular seam. Experiments with ion-exchanged samples of coal have shown that some exchangeable cations in the coal catalyze ignition reactions, and others appear to have an inhibiting effect.

Lithotype identification of the samples studied has established that, within a particular seam, the occurence of poorly igniting coal is confined to the darker "lignitic" bands of coal; the lighter colored "earthy" brown coals have better ignition characteristics. The darker coal also has the higher moisture content (26) so that

Figure 5.30: Effect of Coal Moisture on Pressure Fluctuations in a Combustion Chamber

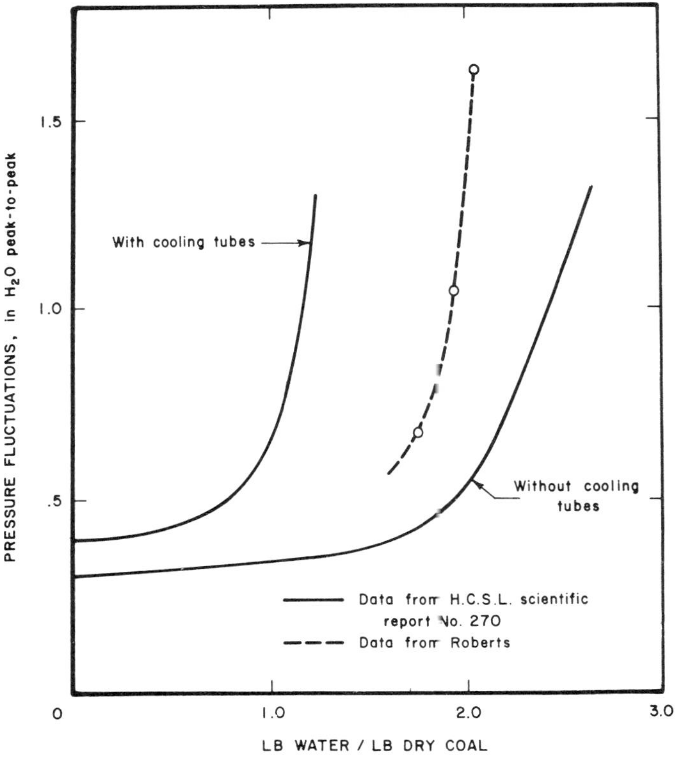

Source: IC 8650

Figure 5.31: Effect of Coal Type on Flame Radiation Temperature

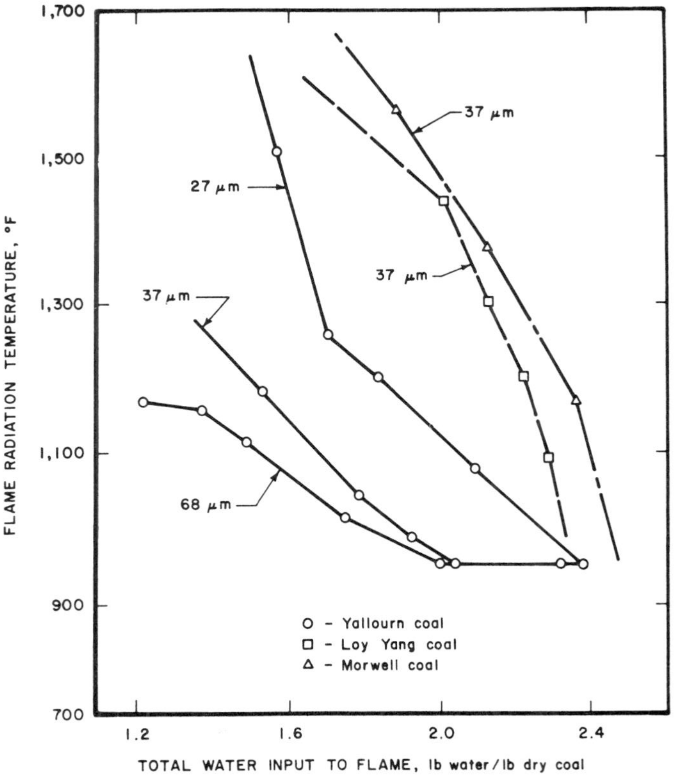

combustion instability problems in Yallourn power stations are likely to be aggravated when the boilers are being supplied with a high proportion of the darker coal. Unfortunately, identification of this banding by color is only possible after the coal face has been exposed for some weeks.

Ash Deposition: Although all seams in the Latrobe Valley area show many similarities, there are important differences in moisture content and inorganic constitution. The opencuts at Yallourn and Morwell are in geologically distinct coal seams.

Morwell coal contains more ash than does Yallourn coal, and this becomes more significant if the coals are compared in terms of effective heat release because of the difference in moisture contents. However, the more important differences are those related to the proportions of the inorganic constituents.

In general, the Morwell coal contains much higher concentrations of calcium and sodium than does Yallourn coal, and the ratio of calcium to magnesium is dominated by calcium rather than magnesium (27). In both coals, calcium, magnesium, and sodium are present as carboxylates, but in the Morwell coal some sodium occurs in the form of sodium chloride. Other elements occasionally present as carboxylates include iron, potassium, and aluminum. Iron and aluminum are often uniformly distributed through the coal matter as oxides or hydroxides. Sulfur is mainly present as part of the coal molecule but also occurs in pyritic form, and chlorine can be attached to the coal molecule or be present as halite (28).

Extraneous ash consists of quartz, clay minerals, and pyritic inclusions (predominantly marcasite), and the amounts present can fluctuate widely. Although cracks filled with sand create local anomalies in the Yallourn coal, the general level of extraneous ash is lower than that in Morwell coal (29)(30).

Ash deposits from combustion of the different coals are of the same sulfate-bonded type, but the differing proportions of the inorganic constituents cause marked differences in the quantities and characteristics of the fireside deposits in the boilers.

The average furnace heat flux in a boiler burning Yallourn coal remains practically unaltered at approximately 25,000 Btu/ft^2 hr after 11 to 12 months of steaming (31). In a boiler fired with Morwell coal, the average furnace heat flux drops from 25,000 to 18,000 Btu/ft^2 hr after 4 to 5 months of operation, and the overall heat transfer coefficient in the secondary superheater decreases from 11 to approximately 5 Btu/ft^2 hr °F (32). In a boiler fired with Yallourn coal, the secondary superheater coefficient drops from 11 to approximately 8 Btu/ft^2 hr °F after 11 months.

Nature of Deposits: The nature of fireside deposits, formed during the combustion of Latrobe Valley coal, has been extensively studied over the last 13 years in programs conducted by the Commonwealth Scientific and Industrial Organization (33)(34) and by the SECV (35).

Gas side deposits found in boilers can be divided into three major groups: (1) Deposits exposed to radiant heat, (2) Deposits formed on convection surfaces, and (3) Loose agglomerates of fly ash.

The first group includes the greatest variety of deposits both in terms of chemical composition and physical properties (36). Samples range from dense siliceous slags to highly sulfated bonded deposits of various degrees of hardness. Slags and clinker-type deposits are found on the sloping walls of the furnace hoppers and around the burner mouths and gas offtake throats. (Figure 5.32 is a schematic diagram of a 200-MW boiler from Hazelwood Power Station.) Hard sintered formations have also been observed in the outer layers of deposits collected from the furnace wall tubes at both burner and gas offtake levels although the inner layers are usually softer and more sulfatic. Slags and clinkers have bulk densities between 1.8 and 2.9 (on occasions as high as 3.1), and the density of the particles can vary from 2.3 to 4.3.

Figure 5.32: Schematic Diagram of a 200-MW Boiler, Hazelwood Power Station

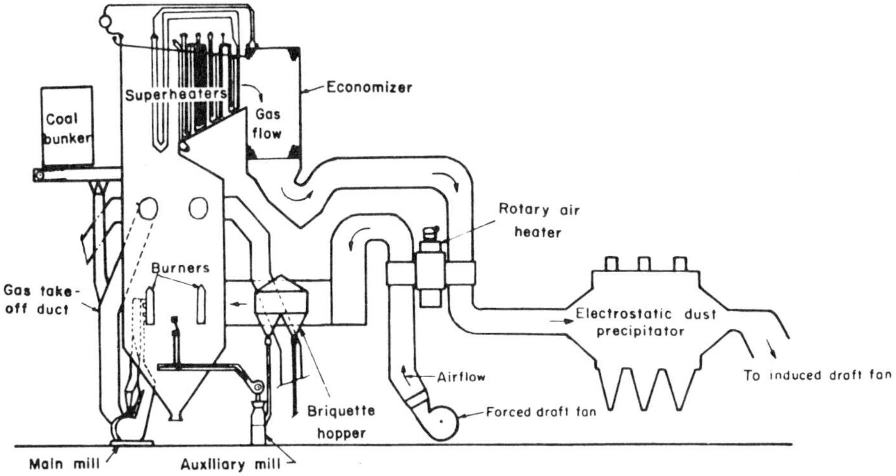

Source: IC 8650

Generally, hopper and burner level deposits are richer in iron and silica than those collected at the gas offtake level (both from water wall tubes and gas offtakes), but there are exceptions.

Sulfatic, bonded deposits are particularly noticeable in the higher regions of the furnace and on platen superheaters. In these areas, they resemble, in composition and physical appearance, the deposits collected from the secondary and primary superheaters.

The individual particles that form hard sintered deposits have a glassy appearance and are translucent. The particles from sulfatic deposits are finer, more crystalline and opaque.

Deposits formed on convection surfaces, such as superheater deposits, vary in bulk density between 1.3 and 2.6 and in particle densities from 2.4 to 3.4. The deposits have a distinct layered structure, and the individual layers themselves are formed by alternating darker and lighter colored bands. The inner layer seldom exceeds 1 to 2 mm (0.040 to 0.080 inch) in thickness. It is soft and consists of submicrometer or 1- to 2-μm crystalline, light colored particles. Chemically, the inner layer is rich in alkalies and sulfates and can contain considerable amounts of chloride. The outer layers are usually darker and can be hard and strongly bonded. The main bonding agents have been identified as thenardite, anhydrite, and mixed sodium/magnesium sulfates.

Examination of deposits by the scanning electron microscope has shown that individual oxide and siliceous particles are enveloped by sulfate and that the envelopes coalesce into a dense sulfatic matrix.

Economizer deposits are not unlike superheater deposits in chemical composition but are usually more friable and are sometimes present in the form of loosely held agglomerates of fly ash. Air-heater deposits are physically similar to economizer deposits but less sulfated.

During a period of boiler operation, the chemical and physical properties of deposits undergo noticeable changes. The main factors responsible for these changes are: (1) continuous sulfation together with some formation of ferrites and silicates; and (2) changes in surface and gas temperatures leading to sintering and smelting, and also to evaporation and the diffusional migration of volatile compounds.

The consolidation of furnace deposits is associated with changes in chemical and mineralogical properties, but similar changes do not accompany the hardening and increase in density of superheater deposits.

An important factor in the consolidation of superheater deposits is migration of sodium salts (37). Examinations by scanning electron microscopy and the electron probe microanalysis of soft superheater deposits reheated in a laboratory muffle furnace have shown the existence of dense sodium-rich layers and the consolidation of the structure when this layer is fused. The so-called outer skin observed on many freshly sampled superheater deposits resembles closely, in composition and appearance, the inner layer and was probably formed on cooling by the migration of sodium salts from the core of the deposit. If fireside deposits are not removed during a shutdown, this sodium-rich outer layer will soften and melt when the boiler is restarted causing a dense layer within the deposit and providing a sticky surface that will retain impacting fly ash particles (38).

In 1966-67, the nature and extent of fireside corrosion of superheater tubes in Latrobe Valley boilers was investigated (39). The observed corrosion rates varied between 0.007 and 0.015 mg/cm^2 hr over an extended period, and generally corrosion of ferritic and austenitic superheater tubes did not appear to be excessive for superheaters conveying steam at temperatures below 1050°F.

Low temperature corrosion has occurred on a number of occasions in air heaters, but when the metal temperatures are maintained above 300°F, the corrosion rate is negligible. Special materials such as vitreous enamel coatings have also been used to improve corrosion resistance at low temperatures.

Boiler Cleaning Practices: Because steam soot blowing was found to be unsuccessful in cleaning deposits from the water tube walls of the boilers at Hazelwood

Power Station, water blowers (water jet deslaggers) were introduced over 6 years ago. Each boiler, generating 1,650,000 lb steam per hr at 1,600 psig and 1055°F, is equipped with 42 water blowers that effectively clean 50% of the furnace area. The boilers are fired with Morwell coal and consume some 6,000 tons of raw coal per day.

There are 38 single-nozzle (4-mm diameter) Babcock & Wilcox water blowers and 4 retractable oscillating water blowers of SECV design installed in the furnace walls and operated in an automatic sequence. Figure 5.33 indicates the locations of all soot blowers on a boiler and gives detailed information on make, model, nozzle size, operating pressures, and speeds.

Each step of the sequence starts three to four blowers and an adjustable timer spaces the steps so that the full sequence is executed during one 8-hr shift; the water blowers situated below the burners are operated twice and the blowers above the burners only once per shift. The delay between the steps of the sequence provides the possibility for manual operation of any blower when deposit buildup in that region requires more frequent blowing. If a consistently stronger cleaning effect is required, the soot blowers can be equipped with heads having two nozzles, each of 4-mm diameter.

Measurements have shown that the heat flux remains practically constant for the surface cleaned by water blower compared with a 75% deterioration in the heat flux after 500 hr for an uncleaned surface.

The tangent tube construction of the boiler furnaces, with no membrane walls and very limited use of finned tubes, was undoubtedly a factor in the successful application of on-load water washing. However, a number of difficulties were experienced during the conversion from steam soot blowing. The long retractable oscillating water blowers, which are installed in the side walls of the furnace hopper and immediately below the nose on the rear wall of the furnace (**141**, **142**, and **139**, **140**, respectively, on Figure 5.33), have proved to be very successful, but have resulted in some water-side circulation problems.

The blower oscillates through 90° and four jets are installed on the head, the two forward-facing jets being intended to clean surfaces past which the blower travels and the two backward-facing jets to clean the vertical side wall through which the blower protrudes. The rear jets also clean the sloping walls of the furnace as the blower advances towards the limit of its travel. Some of the wall tubes that are cleaned by the oscillating water blowers form the roof of the furnace and the superheater enclosure, and it was found that a number of conditions could arise that significantly reduced the natural circulation in these tubes.

Tube temperature measurements have been made from the back of tubes in the furnace roof, and circulating flow rates with tubes have been determined by pitot measurements. It has been found that excessive quantities of water from a blower can reduce natural circulation sufficiently to result in a 45°F increase in tube metal temperature. Blower design changes were made to minimize the excessive cooling when the water jets impinged on the walls near a corner. The simultaneous cleaning of tubes by an oscillating water blower in the superheater area resulted in an increase in tube metal temperature of 135°F and is believed to have to have caused three tube failures. To overcome this problem, the water blower

Figure 5.33: Soot Blower Locations and Equipment Detail as Installed on a 200-MW Boiler, Hazelwood Power Station

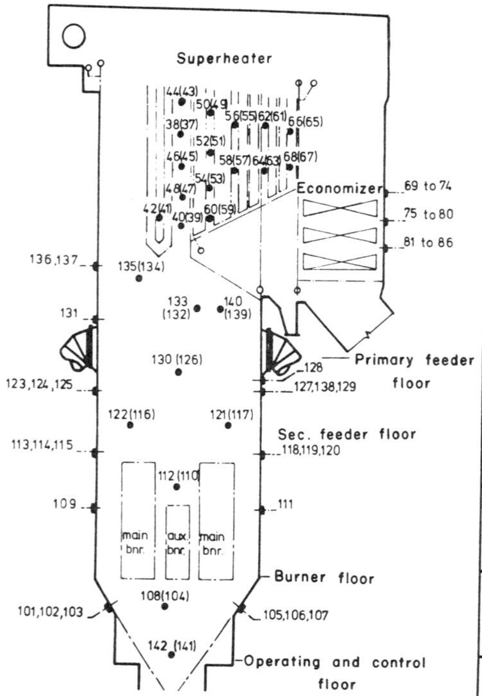

Sootblower type	Blower number	S/B Operating pressure (psi)	Remarks
Furnace (water)	101 to 142	100–180	Babcock & Wilcox: Type, Mk2 Mk3 Travel rate, 6.5 6.5 in/min Rotation, 1.5 3.5 rpm Water consumption, 5 gpm SEC oscillating type: Travel rate, 3 fpm Oscillation, 6/min Minimum distance from tubes, 6 ft Nozzle size, S/B's 101-112, 4-mm diam 113-142, 3-mm diam
Platen (steam)	41 to 42	200	Diamond rack type IK300 25 ft travel, two nozzles 7/8- or 1-in diam
Secondary superheater (steam)	37 to 60	350–380	Diamond rack type IK300 25 ft travel, two nozzles 7/8- or 1-in diam
Primary superheater (steam)	61 to 68	170–200	Diamond rack type IK300 25 ft travel, two nozzles 7/8-in diam
Economizer (steam)	67 to 74 75 to 86	250–280 180	Diamond rack type IK300 20 ft travel, two nozzles 7/8- or 1-in diam

Source: IC 8650

head was redesigned with a 3-mm diameter nozzle, and the automatic control sequence was altered to prevent the simultaneous operation of soot blowers affecting the same tube. The alinement of the two rear nozzles on an oscillating blower so that both jets strike the same tube can stop the natural circulation within that tube, and it is possible that flow could reverse. When monitoring circulation rate on a tube during the operation of an oscillating water blower, a no-flow condition was observed that persisted after the blower completed its operating cycle. The tube metal temperature indicator continued to rise steadily under these conditions until, when 135°F above the usual temperature, the operator reduced load by 25 MW. This sudden load reduction permitted the natural circulation in the tube to reestablish and the tube wall temperature returned to its normal level.

Experience at Hazelwood Power Station has shown that the single rotary water blower (equipped with one or two nozzles) is unlikely to influence the natural circulation in a boiler, even in locations where the heat flux is low, but the simultaneous operation of two adjacent water blowers should be avoided.

Some cracking has been observed on the water-washed sections of tubes in the vicinity of tube butt welds or attachment welds. Accelerated thermal fatigue investigations on tube samples have failed to reproduce this cracking after more than 30,000 cycles, and it is considered that restraint effects must be significant in the boiler. Further laboratory-scale investigations are proceeding, and the operation of the water blowers in the power station is being developed toward minimizing the quenching effect but maintaining effective cleaning of the heat transfer surface.

In the superheater regions, conventional steam soot blowing is used; rack-type Diamond soot blowers with two ⅞-inch diameter nozzles are installed in the platen and secondary and primary superheaters, and are operated in pairs. In the platen and the primary superheaters, steam at a pressure of 200 psig is used, and in the superheater, the steam pressure is 350 psig. Steam supply for the soot blowers is taken from the inlet to the secondary superheater at 1,650 psig and 800°F. The installation on four boilers of a steam pressure reducing station to permit distribution to the soot blowers at 600 psig has considerably reduced maintenance costs. Soot blowing is controlled by an automatic sequencing unit with provision for any soot blower to be operated out of sequence. Normal operation provides for superheater soot blowing twice per shift.

Under typical operating conditions, the flue gas temperature at the inlet to the economizer would rise by approximately 150°F per shift if the superheater soot blowers were not operated. Because the factors currently limiting the length of boiler operation without off-load cleaning are the maximum permissible gas temperature at the inlet to the economizer (1300°F) and the lowest acceptable final steam temperature (1000°F), satisfactory soot blower operation is essential.

A comprehensive system of field reporting and maintenance priorities provides for daily attention to problems not covered by the normal programmed maintenance. The effectiveness of the soot blowers in maintaining the heat transfer capacity of the superheaters is monitored by calculating heat transfer coefficients for the separate superheating stages to determine trends.

Water washing of superheaters is regarded as an interesting but doubtful prospect (40)(41). Laboratory investigations are being conducted on low-alloy superheater steels (1% Cr, ½% Mo and 2¼% Cr, 1% Mo) to determine sensitivity to thermal fatigue. Indications to date are that cracking does not occur in the 1% Cr, ½% Mo alloy at metal temperatures up to 500°C (930°F) and up to 16,000 quench cycles, but there is slight cracking after 30,000 cycles; above 500°C, cracking occurs independent of the number of cycles (42). The two steels investigated exhibit similar tendencies, but the cracking is more severe in the higher alloy steel.

To prevent the overheating of superheater tubes by excessive soot blowing at a time when the furnace heat transfer is restricted, the unit controller is aided by the indication in the control room of metal temperature on 30 selected tubes in the platen and secondary and primary superheaters. During soot blowing, the metal temperature of the tubes cleaned will normally increase by some 45° to 55°F, but this increase will be greater if the steam flow in the tube is low. Consequently, soot blowing is restricted during startup or low-load conditions.

The metal temperature monitoring system installed was developed from temperature measurements made on some 120 tubes and early experience with superheater tube failures. The temperature measurement is made in the boiler vestibule and has been found to be convenient and reliable and also useful as a steam temperature indication. The temperature limits applicable to different groups of tubes are indicated on the temperature recorders, and the operating staff take decisions relating to the extent of soot blowing from the control room.

Efforts are being made to limit the peak in the distribution of gas temperature across the boiler in the region of the superheaters. It has been established that higher temperatures increase the rate of deposition and cause deposits to be denser and more difficult to remove from the boiler during off-load cleaning. The temperature distribution in this region is affected by the combination of mill-burner units that are in service at any time (seven out of eight being required to maintain full load) and by the firing conditions within the furnace (since swirling conditions are generated by the burner system) and is accentuated by the fact that there is no crossover (from the center of the gas path to the side) in the steam flow between the platen and primary superheaters. Some progress has been made in leveling out this temperature distribution by balancing coal feed rate to the mills.

Economizers are of a steaming type and when the boiler is nearing an off-load clean, the steaming rate can exceed 30%. Six rack-type, Diamond steam-soot blowers with two nozzles of ⅞- or 1-inch diameter are provided for each of the three banks of the economizer. The soot blowers are operated in groups of three in sequence from the top bank to the bottom following the superheater soot blowing cycle. The soot blowers mounted above the middle and bottom bank of the economizers are 16 inches or less above the top rows of tubes and erosion has caused tube failures.

Soot blowing was first restricted to once per shift for these lower banks, then on the four boilers equipped with a steam reducing station, the pressure to the soot blowers involved was reduced from 250 to 180 psig. At this pressure, once-per-shift operation provided insufficient cleaning, and the sequence was restored to twice per shift. On the other four boilers with steam pressure at 250 psig, the nozzle diameters were reduced to ½ inch, and the most affected tubes were protected with sleeves. The effectiveness of the different cleaning conditions was monitored by calculating heat transfer coefficients for the economizer from operating conditions and determining the short- and long-term trends.

Shot cleaning was practiced on the tubular air heaters of the first two boilers until vitreous enameled tube ends were installed to reduce acid dewpoint corrosion at the cold tube plant; since then, these air heaters have only been cleaned off-load. The other six boilers have rotary air heaters, each equipped with 12 Babcock & Wilcox, fixed lance, multiport soot blowers and the soot blowing of the air heater is included in the boiler sequence, twice per shift, after the superheater and economizer blowing.

Initial difficulties with high differential pressures across the air heater were caused by the combined effects of plate disintegration and ash deposition. Plate disintegration was found to be caused by vibration under the action of soot blowers using insufficiently superheated steam at unnecessarily high pressures. A reduction in pressure from 250 to 180 psig significantly increased plate life, but tests carried out with an experimental Diamond IK 110 blower have shown that 90 psig is sufficient for effective cleaning so that further improvement might be possible. Acid corrosion conditions exist for a part of each air heating, heat transfer cycle, but vitreous enamel coating of the plates provides good protection.

Boiler operation has improved to the state where off-load cleans are expected after 3,000 to 4,000 hr of operation although abnormalities can force earlier outages. The approach to cleaning is to soak the deposits with water for about 48 hr and then dislodge them with jets of water from hand-held fire nozzles. Some hand cleaning may then be required in the superheater areas but the boiler can be returned to service after 5 to 7 days if no major maintenance is required. The maximum water demand for washing is 70,000 gal/hr, and all the water and ash is sluiced to ash ponds for disposal.

Ash Disposal

Although the ash that forms deposits on the heat transfer surfaces of the boilers creates limiting operational problems, it is only 2 to 4% of the total ash in the coal consumed. The distribution of ash in a boiler at Hazelwood Power Station has been estimated as 85%, collected by electrostatic precipitator, 11 to 13%, collected in furnace hoppers after fallout or removal by on-load cleaning from furnace, superheater, economizer and air heater; and 2 to 4%, collected on heating surfaces as deposit and removed off-load.

Typically, then, burning 4% ash coal in the 200-MW boiler, there will be some 75 to 80 tons/day of fly ash from the precipitators, 10 to 15 tons/day of coarser ash or agglomerates from the various ash hoppers on the boiler and about 2 to 4 tons/day added to the deposits within the boiler.

Electrostatic Precipitation of Fly Ash: Electrostatic precipitators were first installed with the 200-MW units at Hazelwood and are also being installed with the 350-MW units under construction at Yallourn W. The 120-MW units at Yallourn that have been in service for 12 years have been equipped with precipitators, and plans are being developed for the older plant.

Fly Ash Characteristics: The chemical compositions of fly ashes formed from Yallourn and Morwell coals reflect the differences in the original inorganic constituents, but size distributions are very similar. The larger size fractions of fly ash, sampled from the boilers burning Morwell coal, usually contain a considerable proportion of agglomerates formed from light-colored skeletal particles.

Microscopically, the Morwell fly ash consists predominantly of very fine yellow and white crystalline particles, which have been identified as anhydrite, thenardite, periclase, and quartz. It has a relatively low silica content and is less sulfated than most of the deposits. Black magnetic cenospheres ranging in size from 1 mm to less than 66 μm in diameter are frequently observed and consist mainly of magnetite. In contrast, the Yallourn fly ash is less sulfated, has higher contents of iron and alumina, a lower calcium-to-magnesium ratio, is darker, less agglomerated, and contains fewer black magnetic cenospheres.

The essential conclusion from the examination of electron micrographs of freshly formed fly ash particles is that the coal particles burn out to mainly skeletal and cage-like formations, although some inorganic constituents are released as spherical or irregularly shaped particles and the more volatile sodium compounds are released as vapors (43)(44). The volatilized sodium compounds condense readily on cooled metal surfaces and on solid ash particles and are rapidly sulfated. The branches of the skeletal particles consist of oxides of calcium, magnesium, and iron that can react with extraneous ash and form silicates (for example, forsterite) or can be sulfated by the sulfur oxides present in the gas.

Precipitation Efficiency: The physical and chemical properties of Latrobe Valley brown coal ash appear to be favorable for electrostatic precipitation, and in the existing plant, collection efficiencies of well above 95% are achieved. Laboratory trials conducted on coals from different seams show differences in precipitation behavior but indicate that efficiencies of 95% or higher can be expected for the collection of fly ash at future power stations (45).

The main problems encountered in the operation of electrostatic precipitators are corrosion of electrodes and of structural parts, and difficulties experienced in handling the fine, light and fluffy fly ash.

Ash Handling and Disposal: Ash disposal systems used with the boilers burning Latrobe Valley coals consist of hydraulic recirculating systems incorporating an ash settling pond. The greatest operating difficulties encountered with these wet disposal systems are associated with the high water-soluble content of the Latrobe Valley coal ashes: precipitation of salts from solution causing blockages of valves and pipelines; corrosion-erosion of equipment due to the combined effects of the saline solution and entrained solids; disposal of effluent solutions to maintain a stable concentration of dissolved solids in the recirculating system.

The ash from the Hazelwood Power Station, operated on Morwell coal, is sluiced into a settling pond from which the supernatant water containing the soluble salts overflows into a recirculating pond and is then pumped back to a return water reservoir for the cycle to be repeated. A proportion of the water from the ash pond has to be discharged and replaced with fresh general service water to control the salt concentrations in the sluicing water, and a maximum of 500,000 gal/day of saline ash pond water has been diverted. After 15 months, the disposal of this saline effluent to an outfall sewer had to be discontinued because of blockages caused by precipitated calcium carbonate. The saline water can contain up to 900 mg/l of calcium in solution, and the main cause of precipitation was the dissolved carbon dioxide in raw sewage.

Recent investigations indicate that the formation of objectionable calcium carbonate deposits can be overcome by mixing the saline water with the effluent from a local paper mill prior to discharge into the sewer (46). The paper mill effluents contain 1 to 5 g/l of organic matter of a lignin-sulfonate-type, which apparently prevents formation of hard calcium carbonate scale by acting as a dispersant or chelating agent.

Another approach to the alleviation of this problem is by carbonating the saline water with flue gas, and this is being investigated on a pilot scale. Excessive amounts of calcium can be precipitated as calcium carbonate and settled out before the saline water is conveyed to the sewer.

At Hazelwood Power Station, 10 g/l of total dissolved solids is adopted as the concentration limit within the recirculating system itself because extensive deposits of calcium salts are formed in the return water lines at higher concentrations. This limit is equivalent to a conductivity of 14 to 15 mmho/cm.

The fly ash collected by the electrostatic precipitators on boilers fired with Yallourn coal will be sluiced into a large basin that is the worked-out Yallourn North Opencut. The basin holds approximately 2×10^8 gal of relatively pure water (conductivity 400 μmho/cm), which will be recirculated through the ash sluicing system until conductivity rises to 60 mmho/cm.

A solution of Yallourn ash can be concentrated five to six times more than can a Morwell ash solution because it has a lower calcium concentration and a lower pH and is, therefore, less likely to form deposits that can cause blockages in pipelines or valves. The limiting figure of 60 mmho/cm has been selected to insure that sodium sulfate, the main salt present in solution, will not precipitate under the most adverse temperature conditions anticipated.

The main problems facing this disposal system for Yallourn ash are the possible seepage of saline water from the basin that could contaminate ground and surface waters, and the disposal of the effluent after the conductivity of the water in the pond reaches 60 mmho/cm in about 15 years.

Ash Disposal for Future Developments: Alternative systems of ash disposal are being investigated to determine the most satisfactory approach in relation to future power generating developments in the Latrobe Valley. Broadly, the alternatives are the following: recirculating hydraulic ash handling as discussed previously; dry disposal of conditioned ash within overburden dumps or worked-out opencut areas; a once-through hydraulic system with a pipeline carrying ash slurry to the ocean (approximately 60 miles).

For dry ash disposal it will be necessary to minimize dust generation in the handling system. One approach to this is to moisten the fly ash with water after which the ash tends to solidify because of the formation of hydrated sulfates and carbonates of calcium and magnesium. Solidification creates some handling problems, but it may prove to be a significant benefit by reducing the extent of the subsequent leaching of salts from dumped ash. Current investigations cover the physical and chemical properties of water-conditioned ash, its behavior when mixed with different types of overburden, and the practical extent of leaching from ash dumps. The development of equipment suitable for handling and conveying dry or water-conditioned ash has commenced.

An ocean outfall system is an interesting prospect. The water-soluble components of Latrobe Valley ashes are similar to the salts present in seawater so that no serious effects would be expected. However, this system raises many problems of resource management that require serious consideration in addition to problems of solution chemistry such as the precipitation of magnesium hydroxide from Yallourn ash effluent by the admixture of high-pH Morwell ash effluent solution. The feasibility of this system is under consideration.

Although investigated broadly, there would appear to be little prospect for finding a means of utilizing significant quantities of Latrobe Valley fly ash. The large proportion of soluble salts in the ash is the principal limiting characteristic and any treatment process proposed would be additionally complicated by the day-to-day variability in ash composition. No system of treatment and utilization has been devised that can either simplify the problems of ash disposal or contribute to a reduction in ash disposal costs.

REFERENCES

(1) Kinag, K.D., Nack, H., Oxley, J.H., and Reid, W.T., "Fluidized-Bed Combustion of Coals," Paper in *Technology and Use of Lignite Proceedings of a Symposium Sponsored by the U.S. Energy Research and Development Administration and the University of North Dakota, Grand Forks, N. Dak., May 14–15, 1975.* (Comp. by W.R. Kube and G.H. Gronhovd, co-chairmen.) GFERC/IC-75/2, pp. 36–64 (1976).

(2) Ellman, R.C., Belter, J.W., and Dockter, L., *Effects of In-the-Mill Drying on Pulverizing Characteristics of Lignite.* BuMines RI 6074, 18 pp. (1962).

(3) Gronhovd, G.H., Beckering, W., and Tufte, P.H., *Study of Factors Affecting Ash Deposition from Lignite and Other Coals.* Designation 69-WA/CD-1, 9 pp. The American Society of Mechanical Engineers, New York, NY (1969).

(4) Tufte, P.H., and Beckering, W., "A Proposed Mechanism for Ash Fouling Burning Northern Great Plains Lignite." *Journal of Engineering for Power*, pp. 407–412 (July 1975).

(5) Paulson, L.E., and Fowkes, W.W., *Changes in Ash Composition of North Dakota Lignite Treated by Ion Exchange.* BuMines RI 7176, 18 pp. (1968).

(6) Skinner, D.G., *The Fluidized Combustion of Coal.* National Coal Board Research and Development Department (CPC), London (1970).

(7) Majkrzak, R.S., "The Fluidized Bed Combustion of High Alkali Ash-Low Sulfur Western Coals." Unpublished Master's thesis submitted to the University of North Dakota (Dec. 1976).

(8) Gronhovd, G.H., Tufte, P.H., and Selle, S.J., "Some Studies on Stack Emissions from Lignite-Fired Power Plants." *Technology and Use of Lignite. Proceedings: Bureau of Mines-University of North Dakota Symposium, Grand Forks, ND, May 9–10, 1973.* Bureau of Mines IC 8650, pp. 83–102.

(9) Hammons, G.A., and Skopp, A., *NO_x Formation and Control in Fluidized Bed Coal Combustion Processes.* ASME Publication No. 71-WA/APC-3. The American Society of Mechanical Engineers, New York, NY (Dec. 1971).

(10) Evan, T.J., and Kaiser, W.R., *1978 Keystone Coal Industry Manual,* Texas, p. 638.

(11) Goblirsch, G.M., and Sondreal, E.A., "Fluidized Combustion of North Dakota Lignite." The 9th Biannual Lignite Symposium, Grand Forks, ND (May 18–19, 1977).

(12) Levene, H.D., and Hand, J.W., "Sulfur Stays in the Ash When Lignite Burns." *Coal Mining and Processing*, pp. 46–48 (February 1975).
(13) Archer, D.H., Keairns, D.L., Hamm, J.R., Newby, R.A., Yang, W.C., Handman, L.M., and Elikan, L., "Evaluation of the Fluidized-Bed Combustion Process Summary Report." Westinghouse Research Laboratories, Pittsburgh, PA (Nov. 1969–Nov. 1971).
(14) Ehrlich, S., "A Coal Fired Fluidized-Bed Boiler." *Institute of Fuel Symposium Series No. 1: Fluidized Combustion,* International Conference, London, UK, pp. C4-1–C4-10 (September 1975).
(15) Gibbs, B.M., Pereira, F.J., and Beer, J.M., "Coal Combustion and NO Formation in an Experimental Fluidized-Bed." *Institute of Fuel Symposium Series No. 1: Fluidized Combustion*, International Conference, London, UK, pp. D6-1–D6-13 (September 1975).
(16) Jarry, R.L., Anastasia, L.J., Carls, E.L., Jonke, A.A., and Vogel, G.J., "Comparative Emissions of Pollutants During Combustion of Natural Gas and Coal in Fluidized Beds." *Proceedings, Second International Conference on Fluidized-Bed Combustion*, College Corner, OH, pp. I-5-1–I-5-13 (October 1970).
(17) Burke, D.P., "FBC May Be a Better Way to Burn Coal." *Chemical Week* (September 22, 1976).
(18) Mei, J.S., Gall, R.L., and Wilson, J.S., *Fluidized-Bed Combustion Test of Low-Quality Fuels–I. Anthracite Refuse.* MERC/RI-78/1, Morgantown Energy Research Center, Morgantown, WV (May 1977), published March 1978.
(19) Roberts, F.H., "Pulverized Brown Coal for Power Generation in Victoria." *Proc. 2d Conf. on Pulverized Fuel*, London, England, Inst. Fuel, Paper 15, pp. D1–D10 (November 1957).
(20) Kennedy, W.B., and Warren, D.R., *Boiler Furnace Combustion of Raw Brown Coal.* Commonwealth Australia, Dept. of Supply, Aeronautical Res. Lab. Tech. Memo. A.R.L./M.E. 153, 6 pp. (March 1956).
(21) Thomas, G.R., "Ignition of Brown Coal Particles." M. Eng. Sc. Thesis, University of Melbourne, Melbourne, Australia, 138 pp (1970).
(22) Thomas, G.R., "Ignition of Latrobe Valley Brown Coals; Single Particle Studies." State Electricity Commission of Victoria, Sci. Div., Rept. No. 268, 10 pp. (1972).
(23) Anderson, B., "The Effect on Combustion Stability of Water Additions to Pulverized Brown Coal Flames." State Electricity Commission of Victoria, Sci. Div., Rept. No. 270, 7 pp. (1973).
(24) McIntosh, M.J., "The Effect of Drying on the Breakage of Brown Coal." State Electricity Commission of Victoria, Sci. Div., Rept. No. 246, 21 pp. (1971).
(25) McIntosh, M.J., "Prediction of Brown Coal Drying in a Mill-Drying System." State Electricity Commission of Victoria, Sci. Div., Rept. No. 261, 14 pp. (1972).
(26) George, A.M., "Some Physical Properties of Brown Coal Related to Seam and Lithotype." State Electricity Commission of Victoria, Exploration and Geol. Div., Petrological Rept. No. 14, 25 pp (1972).
(27) Durie, R.A., "The Inorganic Constituents in Australian Coals–Part 3, Morwell and Yallourn Brown Coal." *Fuel*, v. 40, p. 407 (1961).
(28) Murray, J.B., "The State of Combination of the Major Ash-Forming Constituents in Latrobe Valley Coal." State Electricity Commission of Victoria, Sci. Div., Rept. No. 145, 5 pp. (1968); No. 203, 8 pp. (1970); No. 224, 5 pp. (1971); No. 233, 3 pp. (1971); No. 249, 4 pp. (1972).
(29) Henderson, P.L. (ed.), *Brown Coal, Its Mining and Utilization.* Melbourne University Press, Melbourne, Australia, pp. 52–56 (1953).
(30) Baragwanath, G.E., "Some Aspects of the Formation and Nature of Brown Coal, and of the Behavior of Brown Coal Ash in Water-Tube Boilers, with Special Reference to Victorian Deposits." *Australasian Inst. Min. & Met., Proc.,* H_2O_2, S. pp. 131–249 (1962).
(31) Waring P.T., and Morley, W.J., "The Measurement of Heat Flux in Furnaces and Its Application to Brown Coal-Fired Boilers." *Inst. of Eng., Australia, Thermodynamics Conf.–Ind. Applications*, Adelaide, Australia, 5 pp. (August 1970).

(32) Hall, T.V., and Bowles, N.F., "Improvement in Availability of Brown Coal Boilers." *Inst. of Eng., Australia, Thermodynamics Conf.–Ind. Applications*, Adelaide, Australia, 8 pp. (August 1970).

(33) Grant, K., and Weymouth, J.H., "The Influence of the Inorganic Constituents in Victorian Brown Coals When Used in Gas Generators and Industrial Boilers." *Inst. of Fuel, Symp. on the Inorganic Constituents of Fuel*, Melbourne, Australia, pp. 127-138 (May 1964).

(34) Brown, H.R., and Swaine, D.J., "Inorganic Constituents in Australian Coals." *Inst. of Fuel, Symp. on the Inorganic Constituents of Fuel*, Melbourne, Australia, pp. 23-44 (May 1964).

(35) Bonafede, G., "The Behavior of Inorganic Constituents During the Combustion of Victorian Brown Coals." *Internat. Conf. on Fireside Corrosion and Fouling in Thermal Power Stations*, Frankfurt/Main, Germany, 17 pp. (May 30-31, 1972).

(36) Bonafede, G., and Dent, M., "Chemistry of Fireside Deposits Formed During Combustion of Latrobe Valley Coals, Parts 1-4." State Electricity Commission of Victoria, Sci. Div., Rept. No. 154, 6 pp. (1969); No. 174, 6 pp. (1969); No. 202, 5 pp. (1970); No. 209, 7 pp. (1970).

(37) Bartlett, D.J., and Bonafede, G., "The Migration of Alkali Salts in an Inert Porous Medium." State Electricity Commission of Victoria, Sci. Div., Rept. No. 178, 6 pp. (1969).

(38) Kiss, L.T., Ingram, L.K., and Herd, K.A., "Preliminary Study of Off-Load Boiler Cleaning at Hazelwoood Power Station." State Electricity Commission of Victoria, Sci. Div., Rept. No. 257, 12 pp. (1972).

(39) Bonafede, G., and Ellery, A.R., "Fireside Corrosion of Superheater Tubes in Latrobe Valley Boilers." State Electricity Commission of Victoria, Sci. Div., Rept. No. P.R.-3, 22 pp. (1968).

(40) Ellery, A.R., "On-Load Water Washing of Superheater Tubes To Improve Boiler Availability, Part 1. Initial Study at Morwell Power Station." State Electricity Commission of Victoria, Sci. Div., Rept. No. 187, 8 pp. (1969).

(41) Newton, J.D., "On-Load Water Washing of Superheater Tubes To Improve Availability, Part 2. Interim Report on the Metallurgical Effects of Two Years Operation in No. 7 Boiler, Morwell Power Station." State Electricity Commission of Victoria, Sci. Div., Rept. No. 230, 5 pp. (1971).

(42) Newton, J.D., "Thermal Fatigue Cracking in 1% Cr ½% Mo Superheater Tube Steels." State Electricity Commission of Victoria, Sci. Div., Rept. No. 265, 9 pp. (1972).

(43) Bonafede, G., and Drew, W.M., "A Study by Electron and Optical Microscopy of Fly Ash Particles From Brown Coal-Fired Boilers, Parts 1 and 2." State Electricity Commission of Victoria, Sci. Div., Rept. No. 181, 5 pp. (1970); No. 215, 5 pp. (1970).

(44) Bonafede, G., and Dimitrakakis, P., "A Study by Electron and Optical Microscopy of Char and Ash Particles Produced During the Rapid Heating of Brown Coal Particles in a Laboratory Furnace." State Electricity Commission of Victoria, Sci. Div., Rept. No. 228, 11 pp. (1971).

(45) Paulson, C.A.J., "Investigation of the Electrostatic Precipitation of Fly Ashes From Loy Yang Seam Coal, Australia." Commonwealth Sci. and Ind. Res. Organization, Miner. Res. Lab., Investigation Rept. No. 90, 7 pp. (1971).

(46) State Electricity Commission of Victoria, Latrobe Valley Department, "Disposal of Saline Water Effluent." Rept. by the Working Group, 5 pp. (February 1972).

LIGNITE GASIFICATION

The material in this chapter has been based on:

Technology and Use of Lignite, Proceedings of the Bureau of Mines-University of North Dakota Symposium, Grand Forks, ND, May 9–10, 1973 (Bureau of Mines Information Circular IC 8650), compiled by G.H. Gronhovd and W.R. Kube of the Grand Forks Energy Research Center for the U.S. Department of the Interior Bureau of Mines, 1974

Technology and Use of Lignite, Proceedings of a Symposium Sponsored by the U.S. Energy Research and Development Administration and the University of North Dakota, Grand Forks, ND, May 14–15, 1975 (NTIS GFERC/IC-75/2), compiled by W.R. Kube and G.H. Gronhovd, Cochairmen, for the U.S. Energy Research and Development Administration, 1975

Technology and Use of Lignite, Proceedings of a Symposium Sponsored by the U.S. Energy Research and Development Administration and the University of North Dakota, Grand Forks, ND, May 18–19, 1977 (NTIS GFERC/IC-77/1), compiled by G.H. Gronhovd and W.R. Kube, Cochairmen, for the U.S. Department of Energy, 1977

Technology and Use of Lignite, Proceedings of the Bureau of Mines-University of North Dakota Symposium, Bismarck, ND, May 12–13, 1971 (Bureau of Mines Information Circular IC 8543), compiled by W.R. Kube and J.L. Elder of Grand Forks Energy Research Center for the U.S. Department of the Interior Bureau of Mines, 1972.

CO_2 ACCEPTOR PROCESS

Introduction

The CO_2 Acceptor Coal Gasification Process is under development by Conoco

Coal Development Company under contract with the Department of Energy and the American Gas Association.

The unique feature of this process is the use of dolomite or limestone to supply heat for the endothermic steam-carbon gasification reaction. Once calcined to the MgO·CaO form and added at high temperature, the dolomite will carbonate exothermally to $MgO{\cdot}CaCO_3$ in the gasifier to provide chemical as well as sensible heat. Its acceptance of CO_2 at the same time serves to reduce the CO_2 level in the gasifier to minimize CO_2 removal in the purification section.

The process is keyed to providing future pipeline-gas requirements for the Midwest, using lignite or subbituminous coal as the starting material.

Process Description

The process utilizes two fluidized bed reactors with flow of both lignite char and acceptor (either dolomite or limestone) between the gasifier and acceptor regenerator. Ground, dried, and preheated lignite coal from the coal storage and grinding area is fed into the gasifier via lockhoppers.

In the gasifier, the fresh lignite enters into a fluid bed of lignite char. Both devolatilization and gasification will be accomplished in a single reactor. Figure 6.1 shows that the gasifier consists of a main vessel section and a smaller boot section. The reason for this configuration is that acceptor is showered down through the fluidized bed of lignite char and is collected in the boot. The higher gas velocity in the boot section is sufficient to fluidize the acceptor and strip out any entrained char. The boot thus provides a reservoir for acceptor withdrawal.

The gasifier is operated at 1500°F and 150 psig. The principal reactions of gasification are the highly endothermic steam-carbon reaction, $C + H_2O = CO + H_2$, and the water gas shift reaction, $CO + H_2O = CO_2 + H_2$. The heat duty required to maintain the desired temperature is supplied by sensible heat of the acceptor, which flows by gravity from the elevated regenerator to the gasifier, and by the exothermic carbonation of the acceptor as it accepts CO_2.

Fresh lignite enters the gasifier from one of two lockhoppers. The point of entry is low in the vessel to obtain maximum vapor retention time for the volatile material from the coal. All of the volatiles are cracked to CH_4, H_2, and oxides of carbon in the gasifier bed. A char bed of about 25 ft is maintained above the transition section of the vessel.

During operation, calcined acceptor enters the gasifier vapor space and showers through the char bed where it reacts with carbon dioxide. Recarbonated acceptor then collects in the boot section from where it is transferred, char free, to the engager pot. Steam, which enters the bottom of the boot as well as at the transition section above the boot, is used both as fluidization gas and as a reactant in gasification. After purification, the gas is a suitable feedstock for methanation. The product of methanation would be pipeline-quality gas.

The recarbonated acceptor is picked up in the engager pot by recycle flue gas or lift gas and pneumatically conveyed 90 ft into the regenerator. Here, a portion of the lignite char from the gasifier is burned with air to produce the heat necessary for recalcination of the acceptor. A fluid bed of acceptor is maintained in the regenerator by a gas velocity high enough to elutriate the very light char ash particles as well as any acceptor formed by attrition. The flue gas leaving the regenerator flows through a cyclone to remove the entrained particulate matter. Further cleaning of this gas is accomplished in a water scrubber. The ash is collected in one of two ash lockhoppers, then slurried to a large settling pond. The acceptor inventory is maintained by continual addition of makeup to the engager pot.

Figure 6.1 indicates use of recycle flue gas as the lift gas for the acceptor. However, a commercial plant would use air. The pilot plant has the facility to substitute air for flue gas to lift the acceptor.

Figure 6.1: Diagram for CO_2 Acceptor Process

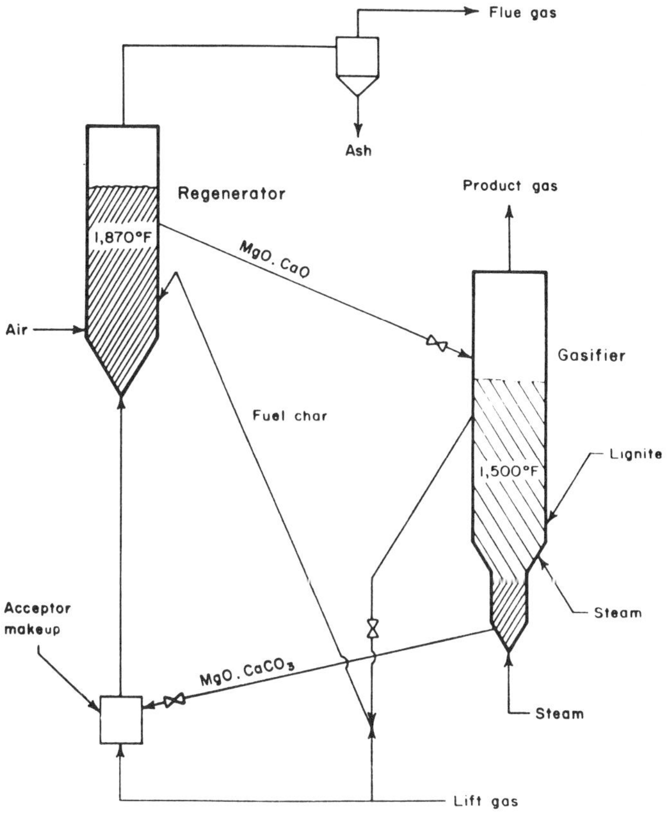

Source: IC 8650

Rapid City Pilot Plant

Stearns-Rogers Corp. built and is operating a pilot plant at Rapid City, South Dakota. The pilot plant at Rapid City has been in operation for ten years.

In the first years, much time and effort was spent in modifying the physical aspects of the plant to make it operative at the unusual conditions required for gasification. The refractory designs in both the gasifier and regenerator were inadequate from the standpoint of physical strength. Both refractory systems failed and replacement required over 3 months downtime. In addition, major design changes were made in the regenerator quench system, the gasifier internal cyclone, the regenerator cyclone, the ash lockhoppers, and the hot line design. Also, the startup procedure has been modified to minimize corrosion of the fired heaters.

Startup Procedure: The startup involves five major steps, each of which has its own potential difficulties. These problems can be overcome by the proper control of one or more key variables. Table 6.1 lists the startup steps, the potential difficulties, and the key process variables associated with each step.

Table 6.1: Startup Sequence

Startup Step	Problems Experienced	Key Process Variables
(1) Acceptor addition and circulation	Pressure balance, loss of seals	Differential pressure between gasifier and regenerator
(2) Char addition to gasifier	Loss of char acceptor interface	Gasifier boot fluidizing velocity Gas distribution Acceptor size consist
(3) Acceptor circulation with char-acceptor interface	Excessive acceptor holdup in gasifier char bed	Char bed density Fluidization velocity Gas distribution
(4) Char combustion to calcine acceptor in regenerator	Deposit formation	Control of regenerator gas composition Fluidizing velocity Gas distribution
(5) Gasification supplying heat of reaction via acceptor	Acceptor transfer line between regenerator and gasifier plugged	Fines content of calcined acceptor Line purges Acceptor circulation rate

Source: IC 8650

The first step involves adding the acceptor inventories in the regenerator and gasifier boot and then circulating the acceptor. The system pressure balance must be maintained to provide solids seals between the gasifier and the regenerator vessels. Each standleg of solids is associated with a maximum allowable pressure drop. Should the system pressure balance require a pressure drop in excess of this maximum, the seal of solids would be lost.

Figure 6.2 illustrates the principle of the solids seal. A pneumatically operated butterfly valve controls the rate of solids transfer through the line. To facilitate transfer, the differential across the valve should be maintained near zero. This is accomplished with a purge gas that is controlled by the differential pressure. The maximum pressure drop in the standleg is proportional to the weight of solids and occurs when the purge gas flow corresponds to the incipient fluidization velocity.

As the regenerator bed is filled, the pressure drop across the lower standleg will increase. By controlling the regenerator pressure at a differential lower than the gasifier, the pressure drops across the two standlegs can be balanced and circulation maintained.

Figure 6.2: Diagram of Solids Transfer Line

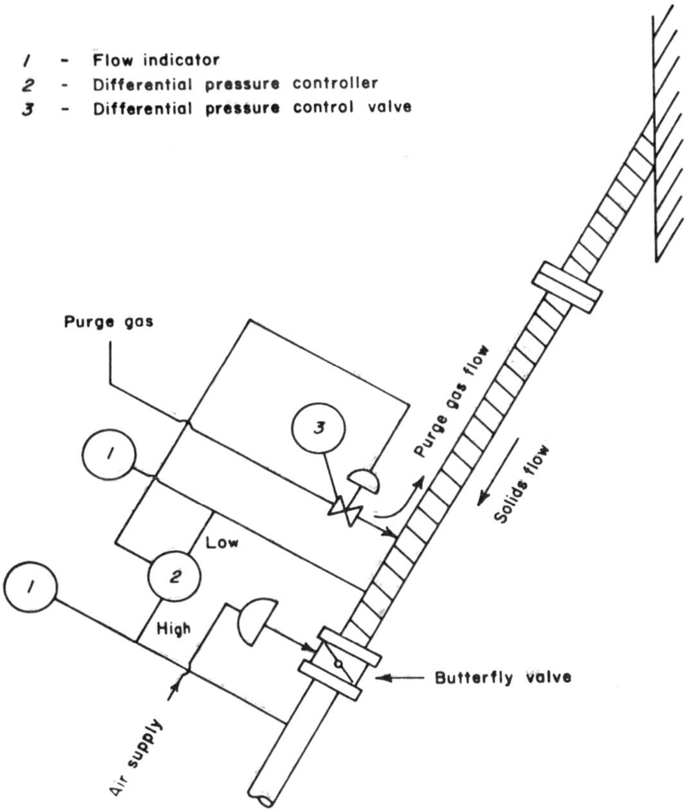

Source: IC 8650

The second step is the addition of char to the gasifier. The major problem in this step is maintaining a separation or interface between the char and acceptor in the boot. Char was added unsuccessfully in the third, fourth, and fifth runs before an interface was maintained in the sixth run.

Two obvious causes for the loss of the interface were the failure of the gasifier refractory and the presence of excessive fines in the acceptor. However, the final solution to the problem was obtained by studying the interface problem in both a glass model and in the gasifier itself.

A stable interface requires operation of the boot at or slightly above the incipient fluidizing velocity of the acceptor. Techniques were developed to experimentally determine the incipient velocity. In addition, a distributor for the boot gas flow was developed. This distributor consists of a vertical closed end pipe with horizontal holes. During the gasifier tests, the interface was maintained after char addition and acceptor was showered batchwise through the bed. These tests were used to devise instrumentation to detect the presence and locate the level of the interface.

In each of the runs made subsequent to these gasifier tests, the acceptor interface has been successfully maintained while feeding the char bed to the gasifier.

Circulation of the acceptor with a char bed present is the third process step. The major difficulty in this step is avoiding an excessive buildup of acceptor in the char bed. The only way of reducing the acceptor holdup is by reducing the char bed density. This is done in two ways: (1) Increasing the fluidizing velocity, and (2) Gasifying the char bed to reduce the density of the char particles.

Since the interface stability requires that the boot flow be near the incipient velocity of the acceptor, the char bed fluidizing velocity must be increased by adding secondary gas above the interface level. The original plant design provided for the addition of secondary gas, but installation of a distributor was necessary to make the flow effective. Another difficulty is that there is no adequate means of directly measuring the expanded char bed density. Experiments made both in the plant and in the glass model show that the differential pressure measurements are insensitive to fluidized bed density changes. The differential pressure measurements are, however, sensitive to changes in the density of the material in the bed.

The particle density of the calcined acceptor is only slightly greater than that of the ungasified char. Therefore, to enable calcined acceptor to shower, the gasifier bed is pregasified by adding air with steam, as the secondary fluidizing gas. Combustion of the char carbon supplies the heat requirement for the carbon-steam reaction. Pregasification was practiced in the seventh and eighth runs and greatly reduced the holdup of acceptor in the bed.

The fourth startup step is char combustion in the regenerator to supply heat for calcination of the regenerator bed. Char combustion has been carried out in the fifth, seventh, and eighth runs.

Significantly, there has been no evidence of ash fusion in the regenerator. However, in the two earlier runs, deposits were formed apparently by a mechanism involving a low-melting CaS-$CaSO_4$ mixture. The reactions involved are as follows: $CaS + 2O_2 = CaSO_4$; $CaS + 4CO_2 = CaSO_4 + 4CO$; $CaS + CaSO_4 =$ Liquid (at temperatures $>1700°F$); and $\frac{1}{4}CaS + \frac{3}{4}CaSO_4 = CaO + SO_2$.

CaS is oxidized to $CaSO_4$ by either oxygen from the air, or CO_2 from the recycle gas. The primary source of CaS entering the regenerator is the char ash. At temperatures above 1700°F, the mixture melts, trapping char ash and acceptor fines. As CaO is formed, the liquid resolidifies leaving a deposit on the reactor wall. The bench-scale work had shown that the presence of CO in the recycle gas suppresses the oxidation to $CaSO_4$. Originally, it was nearly impossible to maintain the desired CO concentration in the recycle gas because irregularities in pressure balance caused erratic fuel char feed.

The fluidization characteristics of the regenerator bed also affect deposit formation. Bed expansions in excess of 50% are required. The gas distribution at the bottom of the bed where the combustion is carried out is also important. To improve fluidization, a ring-type gas distributor was added to the air inlet prior to the eighth run.

During the eighth run, char was burned in the regenerator for about 18 hours. The CO concentration in the recycle gas was kept near the desired level of 4% at all times. Examination of the regenerator bed following shutdown showed that no deposit had formed.

The final start-up step is continuing acceptor circulation with calcined acceptor flowing from the regenerator to the gasifier where it will supply the heat requirements for gasification. During the eighth run, the acceptor return line from the regenerator plugged soon after calcination was complete. Since no agglomerated material was found in the line, the apparent cause for the plug was an excessive amount of fines in the calcined acceptor. These fines tend to be fluidized by the standleg purges. During the run, the transfer of acceptor from the regenerator to the gasifier was momentarily discontinued. This allowed the fines in the standleg to segregate. An accumulation of fine material at any point in the line will prevent the coarser material from moving down the leg.

The calcined acceptor is much weaker than its $CaCO_3$ counterpart, particularly after the first calcination. High attrition rates are expected during first few calcination-recarbonation cycles. On repeated cycling, the acceptor becomes hard and the attrition is expected to drop to essentially zero.

Revised Startup Procedure: In the pilot plant during 1972 and 1973, the initial regenerator inventory for all the runs was fresh dolomite or limestone. With a single fortuitous exception, reliable circulation of calcined acceptor never could be established. Agglomerates or deposits caused by low-melting liquids formed in the regenerator either during the initial calcining of the $CaCO_3$ or shortly afterward. In early operations, control of the regenerator fuel char feed rate was erratic, and the desired level of CO in the off-gas could not be held. Inevitably, the liquids formed as predicted by the bench-scale studies. Later, when better control of the fuel char feed rate was established, agglomerates still formed although not as extensively.

Examination of the cement which held together the agglomerates showed that it contained some finely divided char ash but predominantly was attrited acceptor fines. The bench-scale work showed that: (1) Deposit and agglomerate formation were greatly enhanced by the presence of acceptor fines; (2) the fresh acceptor particles, especially after calcining, have a high rate of attrition; and (3) the acceptor becomes physically rugged and attrition resistant after several recarbonation-calcining cycles.

New information on low-melting liquids in the $CaCO_3$, CaS, $CaSO_4$ system was obtained by laboratory studies at Library, Pennsylvania. With this new information, as well as the above observations, the solution to the start-up problem was now clear:

(1) Eliminate (or minimize the amount of) $CaCO_3$ in the initial regenerator bed while it is being brought up to operating temperature during combustion of fuel char.
(2) Eliminate (or minimize) acceptor attrition.

Dead-burned dolomite (heated to ~3400°F in a rotary kiln) qualifies as the ideal start-up material. It is cheap and readily available, substantially inert to all reactions which occur in the CO_2 Acceptor Process, and compatible with the CO_2 acceptor. It also has a near zero attrition rate.

In the revised start-up procedure, the dead-burned inventory is circulated while the regenerator and gasifier bed temperatures are brought to programmed values. Since the dead-burned dolomite has no activity toward the CO_2 acceptor reaction, the missing heat in the gasifier is supplied by adding air to the steam flow to the distributor ring. Then, fresh stone is added through the make-up feed system. Since most of the regenerator bed is inert material, the instantaneous inventory of $CaCO_3$ is very small, and the low-melting liquids do not form.

Further, since continuous circulation now is assured, the fresh acceptor particles can complete several recarbonation-calcining cycles and thereby develop resistance to attrition. A portion of the recirculating inventory is purposefully withdrawn in proportion to the feed rate of the fresh stone. As the dead-burned inventory is purged from the system, the activity of the circulating burden toward the CO_2 acceptor reaction increases, and the air added to the gasifier is decreased accordingly. Finally, the recirculating acceptor supplies all the gasifier heat duty.

The hours of fresh make-up addition and the hours of operation with no air in the gasifier are tabulated in Table 6.2. At the acceptor make-up rate used, approximately 150 hours are required to totally replace the dead-burned dolomite inventory. Although total replacement occurred only in Run 21, the amount of replacement was sufficient in five of six runs to eliminate the need for air as a gasifier heat source.

Table 6.2: Summary of Plant Operation—1974

Run No.	. . . . .Date		. . . . .Hours.			
	Start	End	At Process Temperatures	With Lignite Feed	With Fresh Acceptor Makeup	Without Air
16	1/14/74	1/23/74	124	Char feed	118	8
17	3/19/74	4/2/74	158	Char feed	72	16
18	4/18/74	4/29/74	156	Char feed	116	0
19	6/5/74	7/19/74	187	150	97	9
20	8/2/74	8/18/74	87	75	53	38
21	8/30/74	9/14/74	251	244	230	171

Source: NTIS GFERC/IC-75/2

Use of Velva Lignite: All of the early demonstration runs made in the pilot plant used precarbonized lignite char as the gasifier feedstock. Char was used to avoid problems which might occur in the gasifier quench system if pitch or tar were formed in the gasifier during periods of upset. In anticipation of smooth plant operation, a special run, Run 15, was made with the gasifier isolated from the regenerator using Velva (North Dakota) lignite as feedstock. During this run, the gasifier heat requirement was supplied by partial combustion with air. The gasifier was operated at 1525°F with a nominal vapor retention time of 30 seconds.

Run 15 was ended voluntarily after 3 days of operation. No process problems were encountered. Results of the run are summarized below:

(1) The dry product gas contained no detectable amounts ($<$0.02 mol %) of any hydrocarbon heavier than methane. Inspection of the quench system and the recycle gas coolers and dryers showed no evidence of tarry matter or of relatively stable materials such as benzene or naphthalene.

(2) There was no evidence of caking or swelling of the char product.

The results of Run 15 led to the use of Velva lignite in process demonstration runs after the ability to operate the plant smoothly was demonstrated in Runs 17 and 18. As shown in Table 6.2, many hours of lignite feed have been logged, and there have been no problems attributable to the use of lignite as feedstock. During Run 21, the gasifier was operated at temperatures as low as 1460°F. At this time, a small but detectable amount of C_2H_6 was present in the gasifier product gas, but there was no evidence of any higher hydrocarbons.

Mathematical Model: In Run 10 and in Run 16, efforts to achieve smooth operation without the use of air in the gasifier were only partially successful. The higher acceptor circulation rates required to supply all the gasifier heat duty generally resulted in an undesirably large showering acceptor inventory in the char bed. The mathematical model, which was developed as a data processing and a process design tool, was used to establish a set of target plant conditions which constituted a new stepwise start-up procedure. This methodical approach was first used in Run 17. In that run, a smooth start-up was accomplished culminating in a 16-hour period in which the acceptor supplied the entire gasifier heat duty. The model also was used to predict the effects of switching from Husky char to Velva lignite as the gasifier feedstock. This was first done in Run 19 with complete success. Thus, it is seen that the mathematical model of the process is useful as an aid to the plant operation.

Solution of Mechanical Problems: Most of the mechanical problems in the plant were solved during the first 2 years of operation. However, mechanical problems continued to limit the duration of the plant runs during 1971. Runs 16, 18, 19, and 20 were terminated by plant failures. These problems have either been solved during the year or will be solved by plant modifications which are currently underway. The key mechanical problems and their solutions are discussed below.

Acceptor Lift Line Erosion – Run 16 was terminated by high temperature areas on the carbon steel shell which contains the alloy acceptor pneumatic transfer line. These hot spots were the result of erosion in bell-shaped sections which comprise the expansion joints of the line. The erosion problem has been solved by: (1) Redesigning the expansion joint to minimize the angle of solids impingement; (2) coating the expansion joint area with Stellite material; and (3) limiting the carrier gas velocity in the line. Since making these modifications, no further problems have occurred.

Fired Heater Corrosion – Sulfur corrosion in the heater coils which preheat gasifier recycle gas was a severe problem during the first two years of operation. This corrosion mechanism was eliminated by installing zinc oxide towers to remove sulfur compounds from the recycle gas. Unfortunately, catastrophic corrosion of the recycle heater coils continued by the carburization-decarburization mechanism referred to as metal dusting. A promising start to Run 17 was aborted due to a heater failure of this nature. Beginning with Run 18, steam was purposefully added to the recycle gas which fluidizes the gasifier acceptor layer. This steam addition has apparently arrested the corrosion problems as shown by ultrasonic thickness measurements and metallographic studies on samples removed from the heaters. Steam serves two roles in preventing attack. First, it preserves protective oxide coating on the surface of the heater coils; and second, it moves the composition of the flowing gas away from a carbon forming condition.

A regenerator heater failure caused the termination of Run 18. This heater coil, which preheats the acceptor lift gas, had been in service since the beginning of plant operation. The failure was shown to be a combination of sulfur and carburization attack. The long and varied history of operation undoubtedly led to the failure. In current operations, the heater is protected from serious attack by the establishment of a maximum operating temperature (1400°F), minimizing the corrosion potential of the recycle gas by avoiding high CO levels, and effective operation of the SO_2 scrubber in the regenerator recycle gas system.

Recent experience has shown that heater corrosion can be effectively controlled in the pilot plant. However, it should be emphasized that these fired heaters are a pilot plant expedient and are not part of the CO_2 Acceptor Process.

Miscellaneous Mechanical Problems — Other mechanical problems which have affected plant operations include stress cracking corrosion and the failure of a fillet weld on the gasifier internal cyclone.

Beginning with Run 19, failures of the expansion joints on the external carbon steel shells of the hot lines have been a problem. A large leak in an expansion joint resulted in the termination of Run 19. Replacement expansion joint bellows were made using a more resistant alloy and were stress relieved. No further problems are anticipated.

Run 20 was terminated when a fillet weld failed causing the gasifier internal cyclone to fall into the char bed. A new weld was made and additional support added to the cyclone which should eliminate further failures.

Intermediate Fines Buildup: During Run 21 with its extended length, a new process problem surfaced. Particles in a size range where they could not be removed from the system built up in the gasifier char bed, affecting the ability to maintain a stable char-acceptor interface. At the end of Run 21, the particles were about 27% by weight of the gasifier char bed.

With the present plant configuration, all material, which is smaller than 65 Tyler mesh, is elutriated from the regenerator fluidized bed. Char ash and attrited acceptor fines fall into this category. Material which is larger than 28 Tyler mesh is removed from the gasifier boot with the reject acceptor. The accumulating fines are in the size range of 28 x 65 mesh. This material circulates by entering the regenerator with the fuel char and returning to the gasifier with the calcined acceptor. Intermediate fines constituted 6% of the calcined acceptor return during Run 21.

These intermediate fines are derived from mineral impurities which exist in the coal seam. These impurities, which principally are α quartz in the Velva lignite, form the nucleus to which acceptor fines adhere. Since the mineral impurities amount to less than 0.2% of the lignite feed, a long run was required to recognize their effect on plant operations.

Solution of this problem involves removal of a portion of the fuel char from the plant. This can be done either continuously or batchwise. In a commercial plant, this purged material could be ground to minus 65 mesh and returned to the regenerator as fuel thus minimizing the thermal penalty involved. A char removal system is currently being installed in the pilot plant. This will allow control of the intermediate fines content of the gasifier bed in future runs.

Future Experiments

Now that the technical feasibility of the process has been proven and most of the mechanical problems have been solved, the experimental program will be aimed at studying the process variables. Pilot plant operation will feature the use of additional coal feedstocks and at least one alternate acceptor. A gas cleanup and methanation system featuring the use of a packed tube reactor is ready for break-in testing. Future runs will test the operation of this system.

Process Variable Studies: The immediate goal of pilot plant operation is an extended run of 30 days duration. This run will feature a series of data producing balance periods at various process conditions. Some of the process variables to be studied include gasifier temperature, regenerator temperature, acceptor makeup rate, gasifier inlet steam partial pressure, and total gasifier steam flow. The data obtained should define the optimum process conditions. If the plant is to be operated for an extended period, the intermediate fines content of the gasifier char bed will have to be effectively controlled using the char removal system.

Alternate Feedstocks: The Velva lignite used as feedstock in recent pilot plant runs has a low sodium content in the ash compared with most North Dakota lignite. The sodium content of the Velva lignite ash is about 2.5% as Na_2O on the sulfur free ash. Lignites with extremely high sodium contents in their ash (14%) were shown to cause gasifier deposits in the bench-scale work.

Glenharold lignite having about 5% Na_2O in its ash has been ordered and will be used as a second lignite feedstock. If this material performs as expected, no sodium related deposits will be formed in the gasifier. Early next year, Glenharold lignite with a high sodium content will be purchased and used as pilot plant feedstock.

The CO_2 Acceptor Process is also adaptable to subbituminous coal feedstocks. It is anticipated that a Montana subbituminous coal will be tested in the Rapid City plant.

Alternate Acceptor: Alternate acceptors will also be tested in future runs. All of the data producing runs made to date have used Tymochtee dolomite from Ohio as the make-up acceptor. Minnekahta limestone, which is quarried near Rapid City, was tested in the bench-scale work and found to be a suitable acceptor. Future runs will test the performance of this stone in the pilot plant.

SLAGGING FIXED-BED GASIFICATION

Introduction

The recent high level of interest in slagging fixed-bed gasification, as evidenced by studies at Westfield, Scotland (1) and the selection by DOE of this process for a demonstration plant, indicates that it may well be the process utilized in second generation gasifiers. The slagging fixed-bed gasifier at GFERC, which had been in standby status since 1965, was reactivated during 1975 and 1976 as a result of accelerated interest in coal gasification. The 24-ton/day unit is the only operable gasifier of its type in the U.S.

The GFERC gasifier is a modified version of the commercially proven dry ash Lurgi fixed-bed gasifier. In the commercial Lurgi process, a bottom grate is used for ash removal and operating temperatures are maintained sufficiently low by excess steam to allow removal of the ash in the dry state. Gasifiers of this type were developed in the late 1930s, and current proposed SNG plants have based their operations on it.

The GFERC gasifier differs from the conventional dry ash Lurgi in that the operating temperatures are sufficiently high to melt the ash for discharge as a slag rather than a dry solid. Two important advantages are obtained in slagging compared to dry ash removal operation. In units of similar size, the steam consumption of a slagging gasifier is about one-fourth of that required for a dry ash unit, while at the same time, the gas production capacity of the slagging gasifier is as much as three or four times that of the dry ash gasifier. Consequently, slagging operation is regarded as a substantial improvement over the present commercial dry ash, fixed-bed gasifier (2).

The slagging gasification pilot plant at GFERC was first designed and operated under the Bureau of Mines, Department of Interior, during the period 1958 to 1965, to determine the feasibility of slagging operation and to assess operational parameters (3)(4). From 1965 until 1974, the pilot plant was essentially maintained in a mothball status. Operation of the GFERC gasifier was reestablished in April of 1976.

The reactivated program at GFERC in slagging fixed-bed gasification has several objectives. The facilities will be utilized to:

(1) Compare the slagging fixed-bed gasification potential and limitations of a wide range of lignites and other Western noncaking coals, as related to feed size, moisture content, and ash composition;

(2) Establish a data base of effluent characteristics, composition and quantity, to provide information to aid in the selection and development of methods for treating and processing these effluents in an environmentally acceptable manner;

(3) Observe and determine the special behavior and characteristics of slags produced from lignites and other Western coals; and

(4) Provide a facility to assist in the development and testing of components or techniques which may expedite commercial-scale gasification.

The size of the unit is well suited for testing a variety of lignites and other Western coals and process modifications. Complete gasifier performance data can be obtained on a single carload lot (60 tons or less), and it is expected that this quantity may also be sufficient to determine extensive effluent characterization data. The pilot plant is supported by the substantial analytical facilities and capabilities of the Grand Forks Energy Research Center.

Process Description

A flowsheet of the GFERC pilot plant facility is shown in Figure 6.3. The gasifier has a maximum throughput of approximately 1 ton/hour of high-moisture (35%) coal. The principal features of the process and test procedure are as follows: test coal, ¾ x ¼-inch size, is charged to the isolated coal lock. The coal lock is then sealed, purged, and repressurized with inert and/or product gas, and the cone valve at the bottom of the lock is opened to allow coal to flow into the gasifier shaft.

As the coal descends in the gasifier shaft, it is first dried and devolatilized by the countercurrent flow of hot gases produced by the gasification of the coal at the bottom of the gasifier. The gasification reaction takes place at approximately 2800°F and is sustained by an oxygen-steam mixture introduced through four water-cooled tuyeres.

During the gasification reaction, the coal is completely consumed, leaving only the molten ash which drains continuously through a water-cooled taphole into a water quench bath. The quenched slag accumulates in the slag lock hopper and is periodically discharged.

The rate of coal flow is governed by the rate at which coal is gasified in the slagging zone. During operation, the coal flow from the coal lock is monitored. When the coal lock empties, it is recharged with raw coal. The time interval between successive empty coal locks is, therefore, the time required to gasify one charge of coal and this data period is used as a basis for calculating the operating performance based on a known quantity of coal.

The raw product gas enters a spray washer and is scrubbed with recycled liquor that has been condensed from the product gas. In addition to the products of the gasification reaction and any entrained coal dust, the raw gas contains water vapor, tars, oils and other organics which are released from the coal during drying and carbonization as it descends into zones of higher temperatures. The major portion of these effluents are removed and accumulate in the spray washer. The washed gas exits the spray washer at 150°F and then enters an indirect gas cooler where it is further cooled to approximately 60°F before it is sampled, depressurized, demisted, metered, and flared.

Some washed and cooled product gas is recycled through the coal lock to prevent any components of the raw gas from condensing in the coal lock and inhibiting the flow of coal into the gasifier. Excess gas liquor from the spray washer and gas cooler is periodically drained into a settling vessel where it is weighed, and sampled, and total effluent quantities are determined.

Figure 6.3: Process Flow Diagram for the Grand Forks Slagging Gasification Pilot Plant

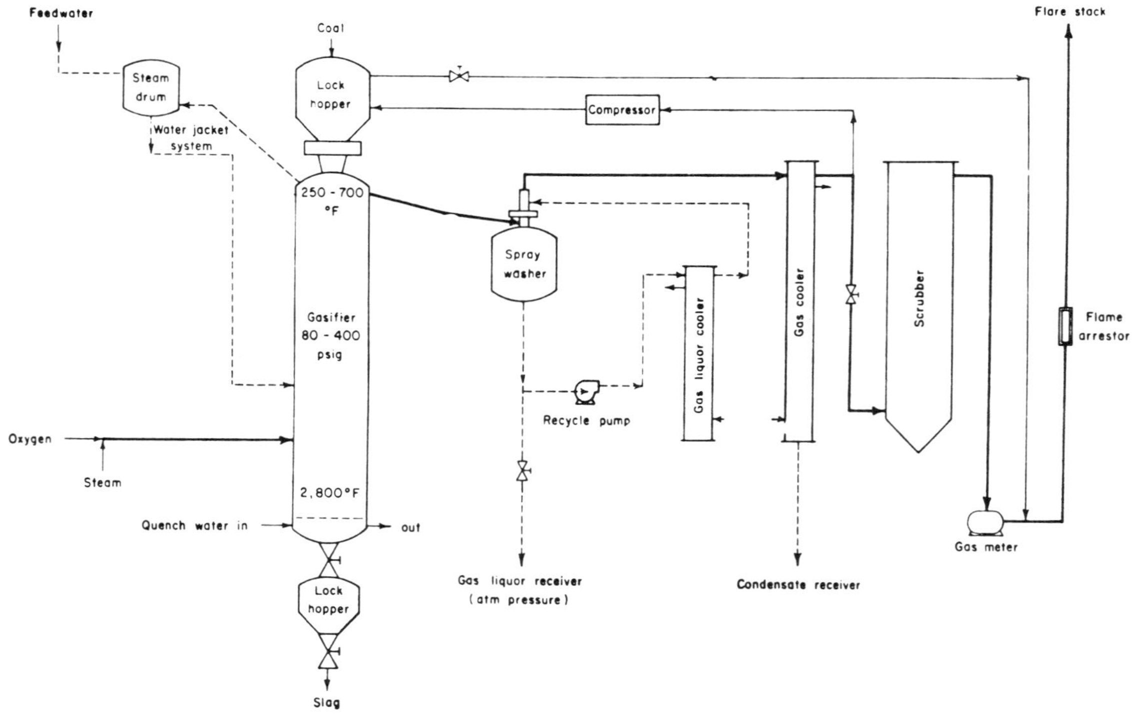

Source: NTIS GFERC/IC-77/1

Design and Specifications of Gasifier: A complete description of the design and initial operation of the system has been published (3)(4). A cross sectional view of the gasifier is shown in Figure 6.4.

Figure 6.4: Cross Section of Slagging Fixed-Bed Gasifier

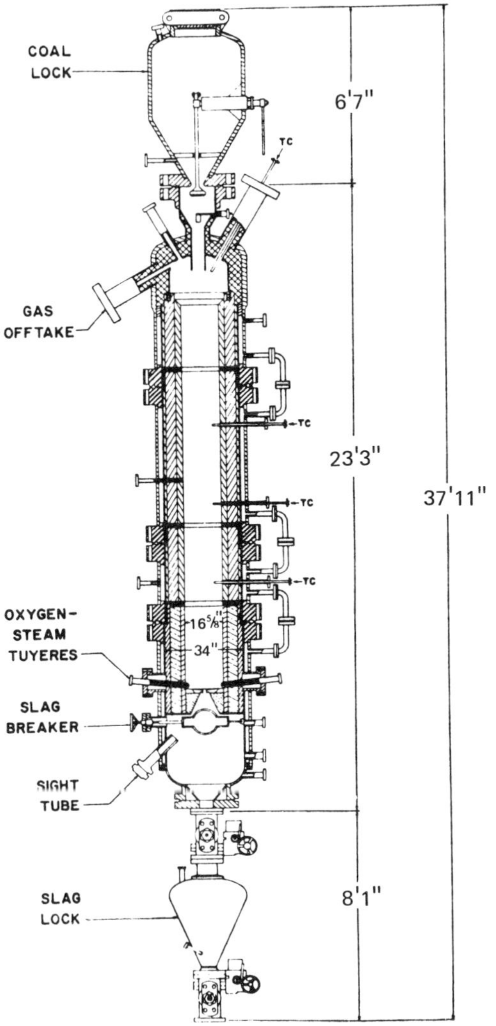

Source: NTIS GFERC/IC-77/1

The overall height of the fuel bed is approximately 15 ft with an inside diameter of 16 ⅝ inches. The design of the gasifier shaft permits the fuel bed height to be altered if desired. By removing either or both of the intermediate sections, the fuel bed height can be changed from 6 to 15 ft. The gasifier shaft is lined with more than 8 inches of refractory. The refractory is backed by a ¼-inch thick liner of 310 stainless steel which forms the inner wall of the water jacket. The outer wall of the water jacket is constructed of 1³⁄₁₆-inch carbon steel plate which forms the load-bearing shell of the gasifier. The pressure shell was designed for a working pressure of 600 psig. The entire top section, with the exception of the connecting flange, is fabricated from 310 stainless steel.

Status of Operational Experience

All tests have been performed with the same single carload lot of lignite obtained from the Indian Head mine of the North American Coal Corp., Mercer County, North Dakota. A typical analysis of the lignite, as tested, is given in Table 6.3. The average moisture content of the coal, as tested (≅26%), was considerably lower than when it was freshly mined (≅35%), due to moisture lost in transportation, sizing, storage, and handling.

Table 6.3: Fuel Analyses for Gasification Tests

Component	Typical Analysis	Range for Coal as Tested
Proximate analysis, %:		
Moisture	25.8	23.9-27.8
Volatile matter	31.1	29.2-32.1
Fixed carbon	34.5	32.3-36.9
Ash	8.6	8.2-10.9
Ultimate analysis, %:		
Hydrogen	6.0	5.8-6.1
Carbon	47.1	43.9-49.51
Nitrogen	0.6	0.6-0.8
Oxygen	37.0	34.9-37.8
Sulfur	0.7	0.5-0.8
Ash	8.6	8.6-10.9
Heating value, Btu/lb	7790	7290-8232
Ash fusion temperatures, °F:		
Initial deformation	1940	1940-2070
Softening	1990	1990-2120
Fluid	2040	2040-2170

Source: NTIS GFERC/IC-77/1

Operating conditions and operating success for shakedown gasification tests conducted since April of 1976 are given in Table 6.4. Of the 16 tests, 6 were completed to scheduled shutdown, 6 were terminated early due to mechanical difficulties associated with auxiliary equipment, and 4 were terminated as a result of problems directly related to operation. The majority of causes for shutdown were attributed to the normal debugging which accompanies use of equipment which has not been operated for years, and, in one case, loss of electrical power in a wide area of the city.

Typical material and heat balances are given in Tables 6.5 and 6.6. The material balance shows an overall unaccounted for value of 1.3%.

Table 6.4: Summary of Gasifier Operation for Initial Shakedown Phase

Run No.	Operating Pressure (psig)	Oxygen Rate (scfh)	Oxygen-Steam Ratio (mol-mol)	Slagging Operation (hr)	Reason for Shutdown Mechanical	Operational	End of Test	Comments
RA-1	80	2,000	0.9	0.6	x	–	–	High O_2 in bath section
RA-2	80	2,000	0.9	–	x	–	–	Electrical power failure
RA-3	80	2,500	0.9	4.1	–	–	x	Taphole coil burned out
RA-4	200	4,000	0.9	3.5	–	–	x	Taphole coil burned out, loss of fuel bed
RA-5	200	4,000	0.9	0.8	–	x	–	Lost slag flow
RA-6	200	4,000	0.9	–	x	–	–	Damaged cone valve at coal lock
RA-7	200	4,000	0.9	4.4	–	–	x	Boiler control problems, steam flow questionable
RA-8	200	4,000	0.9	5.1	x	–	–	Boiler control problems at end of test
RA-9	200	4,000	0.9	1.6	x	–	–	Boiler control problems
RA-10	200	4,000	1.0	0.4	x	–	–	"Free flag" mechanism became disassembled
RA-11	200	4,000	1.0	4.7	–	–	x	–
RA-12	200	4,000	1.0	6.9	–	–	x	–
RA-13	400	4,000	1.0	1.4	–	x	–	Lost slag flow
RA-14	400	4,000	1.0	3.6	–	x	–	High fuel bed loss
RA-15	400	4,000	1.0	–	–	x	–	No slag flow
RA-16	400	4,000	1.0	7.4	–	–	x	–

Source: NTIS GFERC/IC-77/1

Table 6.5: Typical Material Balance (RA-12); Data Period: 6.73 Hours

	Pounds	Pounds per Hour	Percent
Material in:			
Steam	1,279	190	11.2
Oxygen	2,274	338	20.0
Fuel	7,500	1,114	65.8
Purge gas	339	50	3.0
Total	11,392	1,692	100.0
Material out:			
Product gas	8,205	1,219	72.0
Slag	500	74	4.4
Condensate	2,537	377	22.3
Unaccounted for	150	22	1.3
Total	11,392	1,692	100.0

Source: NTIS GFERC/IC-77/1

Table 6.6: Typical Heat Balance (RA-12); Data Period: 1 Hour, Reference Temperature: 60°F

	M Btu	Percent
Heat in:		
Fuel	8,675	97.0

(continued)

Table 6.6: (continued)

	M Btu	Percent
Steam	237	2.6
Oxygen	32	0.4
Total	8,944	100.0
Heat out:		
Product gas:		
Potential	7,306	81.7
Sensible	90	1.0
Slag	69	0.8
Water vapor	357	4.0
Tar (potential)	417	4.6
Heat loss	705	7.9
Total	8,944	100.0

Source: NTIS GFERC/IC-77/1

Operational Considerations

Slag Characteristics: The ash fusion temperature, chemical composition, amount and viscosity of the ash in the coal are important considerations when a coal is utilized for slagging operation. Indian Head test lignite (fluid temperature of the ash <2200°F) has performed well in slagging operation. Tests at 80 psig (2,500 scfh O_2 rate), required additional ash in the form of clinkers (produced from combustion of Indian Head lignite) to be added to the lignite feed since the gasification rate was too low to produce sufficient slag from the lignite to sustain a steady flow through the taphole. All tests at 200 and 400 psig were conducted without the addition of supplemental ash.

Oxygen-Steam Ratio: The oxygen-steam molar ratio has a great effect on slagging conditions, since too low a ratio will favor the endothermic steam reaction and supply insufficient heat to maintain slag flow. Too high a ratio will produce too much heat, which can vaporize components of the ash and cause a bridge to form in the fuel bed when they condense in the cooler upper zones. The range of oxygen-steam ratio which has been found to be acceptable in slagging gasification is approximately 0.7 to 1.2 (2).

However, the range for most lignites (based on past experience in the GFERC gasifier) is more limited, 0.9 to 1.1. The first 8 gasifier tests were performed at an oxygen-steam mol ratio of 0.9. For subsequent tests, the ratio was increased to 1.0. Operation is satisfactory at both ratios; however, the higher ratio produces higher temperatures in the slagging section which improve slag flow characteristics.

The difficulty in maintaining slag flow with too low an oxygen-steam ratio was demonstrated in tests RA-7 and 9 and at the end of RA-8. Due to a sporadic malfunction in the boiler condensate trap, liquid water was discharged along with steam from the boiler. As a result, actual steam flow was much higher than indicated, and, in all 3 tests, problems with slag flow caused termination of the tests.

Oxygen-Steam Rate: With the present test coal, operating pressures, and molar ratios used to date, the oxygen-steam rate has been found to be the major factor controlling the rate of coal gasification. In early tests at 80 psig, a low oxygen-steam rate was employed to ensure that fuel bed hangup did not occur because of excessive gas velocities at low pressures and high gas production rates. In later tests, a higher rate was selected which could be used for both 200 and 400 psig tests. By using the same oxygen-steam rate at both pressures, it was possible to isolate and study the effect of pressure on gasifier effluents without the added variable of gas production rate. Greater gas production rates could be easily achieved in the GFERC gasifier by using higher oxygen-steam rates (4). Tests at higher steam-oxygen rates will be performed in the future.

Taphole Erosion: The most critical problem in maintaining sustained slag flow is to preserve the integrity of the taphole. The high temperatures (≅2500°F) and erosive characteristics of many coal slags create a very severe environment for the hearth plate and taphole. The problem is further compounded by the reduction of iron oxide in the fuel ash to molten iron, which then separates from the slag and can attack hearth plate materials.

Stability of Operation: Following ignition of fuel, a preheat period of approximately 4 hours precedes the appearance of the first slag through the taphole. During this time, the temperatures of the refractory and fuel bed increase to operating status. The temperature of the refractory continues to increase during the first portion of slagging operation and then levels out. In Figure 6.5, internal temperatures are plotted versus accumulative operating time for 2 gasification tests in which operating pressure was the only variable.

Figure 6.5: Fuel Bed and Refractory Temperatures During Tests RA-12 and RA-16

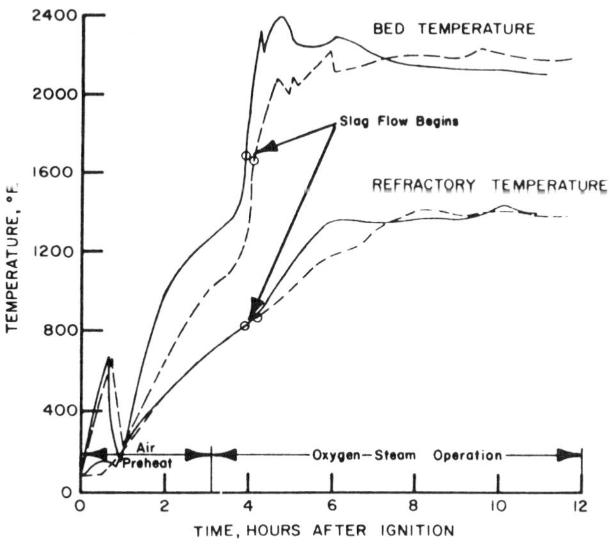

Source: NTIS GFERC/IC-77/1

The upper two curves show gasifier bed temperatures at a level 5½ ft above the hearth plate and about 1 to 2 inches into the fuel bed, while the lower curves show temperatures in the refractory behind the inner layer of silicon carbide tile. Bed temperatures leveled out approximately 5 hours after ignition (1 hour after first slag flow), while refractory temperatures reached a steady condition 6 to 8 hours after ignition 1.

Reproducibility: The ability to reproduce test results at a given operating condition was verified during test RA-11 and RA-12. The slight differences in results for test RA-11 compared to RA-12 are largely due to normal variations in feed coal quality and approach closely the precision of measurement of the operating parameters. Average product gas offtake temperatures were within 10°F, while the maximum variation was less than 4%. Fuel rates closely paralleled each other, with a maximum difference of about 35 lb/hr or 4½%.

Effect of Pressure on Gasification Parameters: To provide a valid comparison of results at two different operating pressures, operating conditions for RA-16 were identical to those for RA-11 and RA-12, except that the operating pressure was increased to 400 psig. Gasifier performance is closely comparable at the two pressures. The fuel rate, the product gas rate, and oxygen and steam consumption values based on the product gas rate showed little difference with pressure. However, the oxygen and steam requirements, based on product gas CO + H_2 production, were somewhat higher at the higher pressure due to slight changes in product gas composition. Cold gas and operational efficiencies also appear to increase somewhat with an increase in pressure.

The individual components of the product gas stream (as sampled under steady state conditions midway between coal chargings) at 200 and 400 psig are compared in Figure 6.6. At 400 psig, the CO_2 and CH_4 content was slightly higher and the CO and H_2 content was slightly lower than that produced at 200 psig. However, pressure as a variable does not appear to have significant effect on gas composition for the range tested.

Figure 6.6: Gas Compositions at 200 and 400 psig

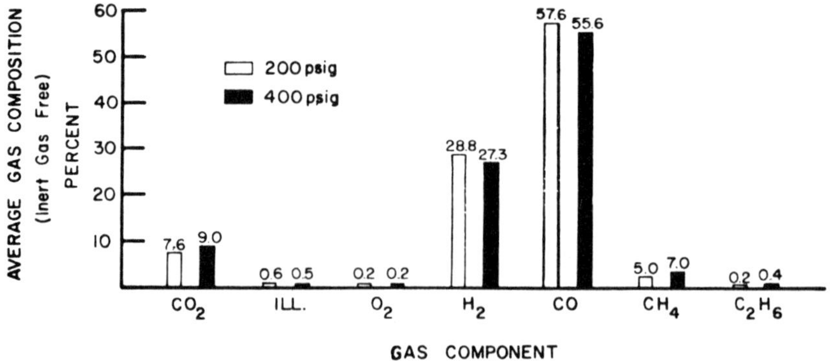

Source: NTIS GFERC/IC-77/1

For 200 and 400 psig operating pressures, gas composition and offtake temperatures at 5 minute intervals for one data period are plotted in Figures 6.7 and 6.8.

Figure 6.7: Gas Composition and Gas Offtake Temperatures During One Data Period—200 psig

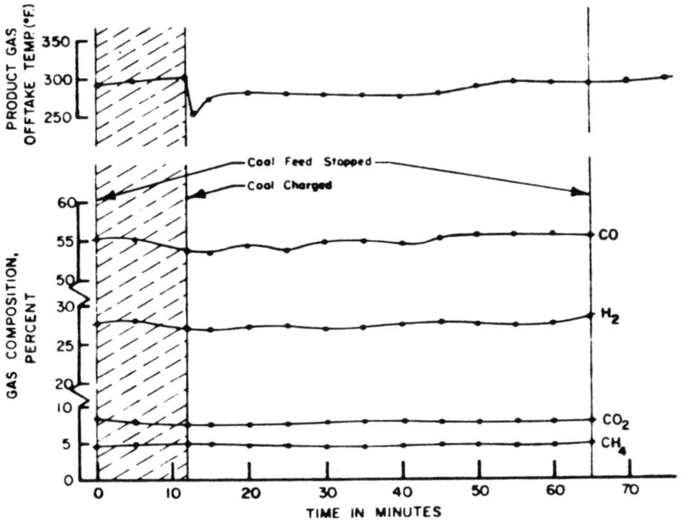

Figure 6.8: Gas Composition and Gas Offtake Temperatures During One Data Period—400 psig

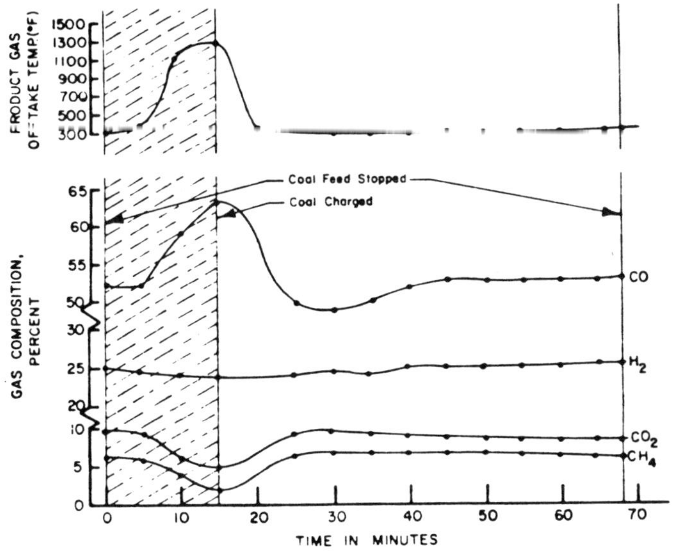

Source: NTIS GFERC/IC-77/1

Moisture content of feed in both tests was similar (≅25%). At 200 psig, there is little change in the gas offtake temperature or product gas composition when the coal lock is being recharged. After fresh coal flows into the gasifier, the temperature drops approximately 55°F and then returns to a relatively stable temperature for the remainder of the data period. The temperature fluctuation is accompanied by no significant changes in gas composition. At 400 psig, there is an appreciable effect during coal lock charging on gas offtake temperatures and gas composition.

During the interval that fresh coal is entering the gasifier, upper bed temperatures are greater during 400 psig operation. As a result, when the fresh coal flow stops and the charging cycle begins, there is less volatile matter to be vaporized from the coal and temperatures increase more rapidly. Product gas offtake temperature increases from just over 300°F to a maximum of over 1300°F before the fresh coal flows into the gasifier. The significant temperature increase favors the formation of CO at the expense of CO_2 and CH_4, while hydrogen concentration shows only a slight decrease.

Future Program

The objectives and status of the reactivated program in slagging fixed-bed gasification of Western coals has been described. In a very recent action, a proposal which will significantly expand the capabilities and research objectives of the GFERC gasifier program has been approved. In addition to upgrading the pilot plant auxiliaries, operational improvements include the addition of a second coal lock to stabilize operation and the installation of a stirrer in the upper portion of the gasifier bed to permit operation on agglomerating coals. As a result, the pilot plant can contribute data and information in support of the demo-scale plant program.

HYGAS PROCESS

Introduction

The plans for U.S. energy independence have stimulated the development of coal gasification technology by which the rich, indigenous coal resource can be converted into clean-burning, economically transportable synthetic gas to keep the nation's pipelines full. Although commercial technology is available today, a new generation of processes, which offers a significant increase in conversion efficiency and a reduction in capital and operating costs, is being intensively developed.

Included in the joint government-gas industry program to accelerate the commercialization of coal gasification is the Institute of Gas Technology Hygas project. Funding is handled through the Department of Energy and the American Gas Association. The throughput of the Hygas pilot plant gasifier is 80 tons of coal per day, while the commercial-size reactor would have a capacity of 5,000 to 7,000 tons of coal per day. The details of the Hygas Process have been described in other publications (5)(6). The discussion that follows will point out the significant operating results from the testing of lignite.

During the first phase of the Hygas pilot plant program, the hydrogen for the gasification was supplied by steam-reforming of natural gas in a package hydrogen plant. The second phase of the program involves the testing of a steam-oxygen gasifier for production of hydrogen from the hydrogasified char. During the first phase of the program, 27 tests were conducted, during which close to 3,000 tons of Montana lignite were processed.

The entire plant, from coal preparation through catalytic methanation, has been operated as a complete unit as shown in Figure 6.9 (7). Pipeline-quality gas was produced with a heating value of up to 1,000 Btu/scf on a nitrogen-free basis. All tests since mid-1972 were carried out at 1,000 psig pressure. So far in the second phase of the program, 6 tests have been conducted using steam-oxygen gasification to produce the hydrogen.

In a recently completed test, the entire integrated plant, from coal feed to methanation, was operated as shown in Figure 6.10 (8). The steam-oxygen gasifier performed well, and the only discharge from the gasifier was ash residue. Thus, the technical feasibility of the Hygas Process combined with steam-oxygen gasification was demonstrated on a pilot plant scale.

Figure 6.9: Sections of Hygas Plant in First Phase of Operation

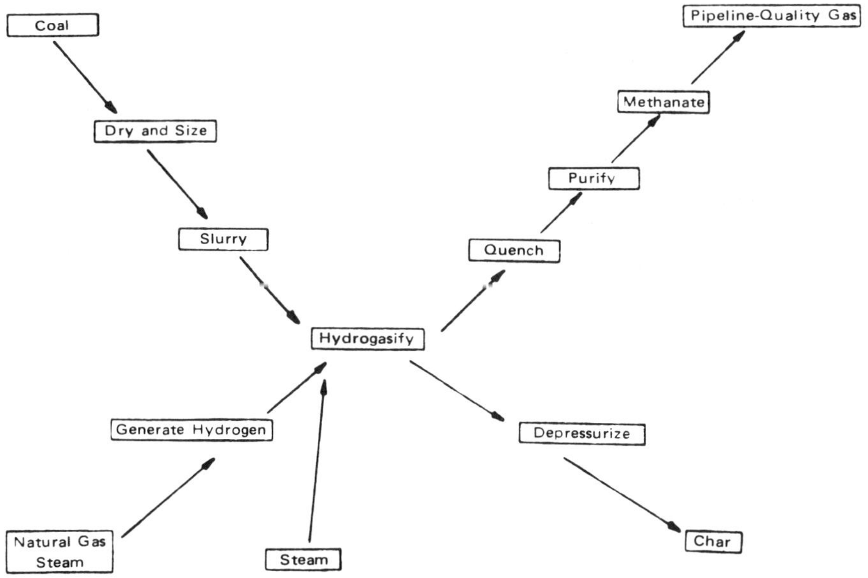

Source: NTIS GFERC/IC-75/2

Figure 6.10: IGT Hygas Process (1)

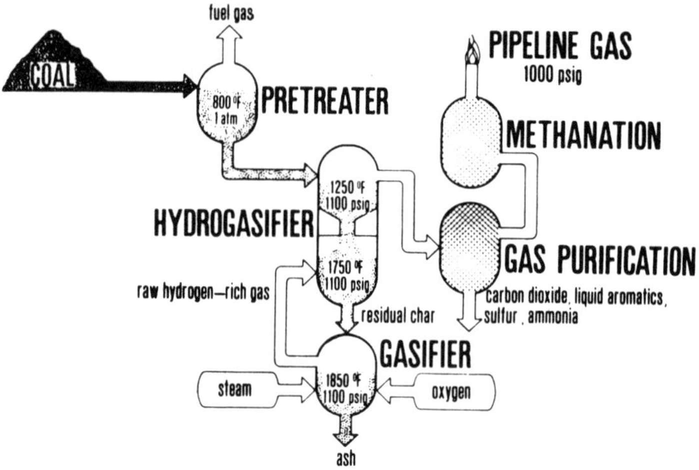

Source: NTIS GFERC/IC-75/2

Lignite Handling

The lignite used at the pilot plant comes from the Savage mine in Montana. It arrives at the plant by rail and has a size of 1¾ x 0 inches. After drying and crushing, the coal typically has the chemical analysis and size distribution given in Table 6.7.

Data on the as-received lignite are also included for comparison. Note the high moisture content, about 34%, of the as-received lignite. To reduce the moisture to below 10% requires a large drying capacity in the coal mill. Also, note the significant content, about 4.6%, of fine, -100 mesh (150 μ) particles in the as-received lignite. Therefore, the incremental fines generation in the mill must be carefully controlled to avoid feeding too fine a material into the gasifier's fluidized beds. Adjustment of the mill speed and removal of half of the original six journals in the mill have been necessary to reduce the amount of fines generated.

In the slurry feed section, the centrifugal pumps with wetted parts made of cast Ni-Hard have been in service for over 1,000 hours without excessive wear. The pumps are belt-driven at 1,550 rpm and circulate a coal-oil slurry of up to 45% coal by weight. The high-pressure, Wilson-Snyder reciprocating pumps, using check valves and seat assemblies made of stainless steel with stellite hard-face, can operate well over 1,000 hours without difficulty.

In the slurry discharge section, the centrifugal pumps with high-pressure casing provide adequate service for circulation of a char-water slurry of up to 25% char by weight. A stellite inlay coats the impeller and the casing and provides excellent protection against erosion. The cast Ni-Hard casing, available for low-pressure service at a low temperature, is not acceptable at high pressure and 450°F because Ni-Hard is brittle.

The discharge slurry pressure is reduced from 1,000 psig to atmospheric pressure across a single choke, which is commonly used to let down oil well drilling mud. The two tungsten carbide faces rotate with respect to each other to control the flow opening.

Table 6.7: Montana Lignite Analysis

	Gasifier Feed (After Drying and Crushing)	As Received
Proximate analysis (wet), wt %		
Moisture	4.05	33.93
Ash	12.66	
Volatile matter	40.17	
Fixed carbon	43.12	
Sulfur	1.04	
Ultimate analysis (dry), wt %		
Carbon	65.24	
Hydrogen	4.47	
Nitrogen	1.16	
Chlorine	0.04	
Sulfur	1.08	
Ash	13.19	
Oxygen	14.82	
Size distribution, USS		
2½"+8	0	67.4
-8+12	0.9	10.6
+20	16.8	7.4
+30	12.8	2.9
+40	13.6	2.2
+60	19.7	2.9
+80	4.8	0.6
+100	9.3	1.4
+200	11.2	1.7
+325	5.1	1.1
Pan	5.8	1.8
Bulk density, lb/ft^3	48.8	

Source: NTIS GFERC/IC-75/2

Gasification

The gasifier reactor section is shown in Figure 6.11. So far, the major process problem that was resolved in the pilot plant program was that of the limitation to solids flow in the gasifier transfer system. After many attempts and many modifications of the reactor piping internals and of the aeration arrangements, experimental proof finally established that a circulating load of fines in the upper stages of the Hygas reactor contributed to the limitation of coal feed capacity and to the inability to maintain pressure balance and flow control. By reducing the cross section of the slurry dryer freeboard space so that it is the same as the bed itself, fines in the bed can leave the system and be collected by an external, high-pressure cyclone. This arrangement has been in operation since December 1973.

Figure 6.11: IGT Pilot Plant Hydrogasification Reactor Sections

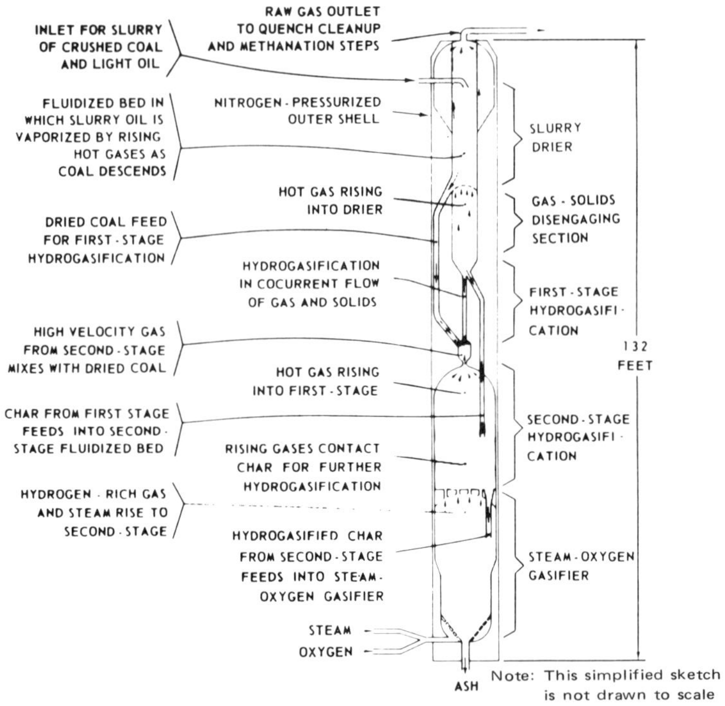

Source: NTIS GFERC/IC-75/2

In May 1974, the lower section of the original gasifier reactor was converted to the steam-oxygen gasifier stage. That space, previously used as a heat-exchange bed, was allotted in the original design for just such a need. The steam-oxygen gasifier is a fluidized bed that operates at up to 1900°F without slagging of the ash. The feed gas is a mixture of steam and oxygen distributed across a specially designed ring distributor with individual cones. Prior to converting the steam-oxygen gasifier, a 27-day test was conducted (Test 27) using hydrogen supplied by a package hydrogen plant. After the conversion, a number of shakedown tests were conducted, culminating in the recent Test 33 wherein the operability of the entire combined Hygas/steam-oxygen system was demonstrated on a pilot plant scale in a fully integrated mode.

Since integrating the steam-oxygen gasifier, the overall carbon gasification has been very high. The residue is almost all ash, indicating essentially complete gasification without requiring the high temperature for slagging the ash. The design oxygen rate is 0.25 lb/lb dry coal, and the steam/oxygen ratio is 4.8 lb/lb. It was also noted that much of the fines collected in the cyclone contained high-ash material, indicating that a good part of the highly gasified residue leaves the system overhead rather than at the bottom. This is consistent with the experience noted in the CO_2 Acceptor plant, where the ash is removed overhead also. With a coal other than lignite, the situation may change.

Gas Purification

The gasified effluent is quenched with water to recover oil for recycle and to remove unreacted steam in the form of condensate. The separation of oil and water in the presence of fine dust has created difficulties. By adding a prequench tower with provisions for blowing down the mud that collects at the interface between oil and water, the quench system can be made to operate satisfactorily. To complete the oil recovery, a stripping system is being constructed that would then steam-strip any blowdown or water discharged from the quench system to recover the oil for complete material balance.

After the quench, the pilot plant purification system consists of a Diglycolamine-water absorber, followed by a caustic wash and a water wash. Not only is carbon dioxide removal complete, but the sulfur removal is essentially complete in one stage of amine absorption.

The caustic wash merely acts as a guard chamber to protect the methanation catalyst. The purification system has withstood pressure upsets, as well as flow changes, and still produces purified gas with less than 0.1 ppm of catalyst.

The use of a caustic wash as a guard against sulfur is convenient for the pilot plant and, perhaps, even for a commercial plant. There is no solid adsorbent, such as activated carbon, or solid reactant, such as iron oxide or zinc oxide, to remove the last trace of sulfur. Such systems are to be avoided because the installation of two parallel units is necessary to periodically regenerate or replace the solids.

There has been some difficulty with foaming in the amine system, which was first abated by the addition of a polyglycol additive. This worked for a period of time, but then lost its effectiveness. Recently, an activated carbon filter, operating continuously on a slipstream, was installed that completely eliminated the foaming problem.

Catalytic Methanation

The methanators were placed in service in April 1973 and have accumulated 630 hours of operation. The details of the performance of such a unit are described by Bair, Leppin, and Lee (8). The packed-bed downflow reactors are loaded with ¼-inch pellets of Harshaw Chemical Ni-0104T nickel-on-kieselguhr catalyst. The heat release is controlled by the IGT split-flow cold gas recycle method.

The conversion of carbon monoxide is complete in all tests. However, because there is no water-gas shift reactor in the Hygas pilot plant due to a funding limitation at the time the plant was constructed, the hydrogen-carbon monoxide ratio cannot always be adjusted before methanation. If the hydrogen-carbon monoxide ratio is less than 3.1:1, hydrogen from the package hydrogen plant is added. However, if the ratio is greater than 3.1:1, no adjustment is made, and the methanation product gas has excess hydrogen, which in turn reduces the gas heating value. The high ratio does not affect the methanator's performance. In a commercial plant design, though, a shift converter section is always included.

The temperature control in the methanators is excellent. The temperature in the methanators increases from 550°F at the inlet to less than 900°F at the outlet. No hot spots have been noted anywhere in these two beds. The temperature front is flat, and the radial profile is uniform. A space velocity of over 4,000 hr^{-1} has allowed complete conversion using less than one-third of the catalyst bed. The cold gas recycle method of temperature control is effective, as well as economical in operation. The overall recycle ratio is typically 1.5 parts of recycle gas/part of feed gas.

Commercial Plants

For a full-size commercial plant producing 250 million scf per day of pipeline-quality gas, three parallel gasifier trains are visualized in the plant. The daily coal consumption would be about 27,000 tons of as-mined lignite. The water requirement would be around 5,000 gal/min, when air cooling is maximized. This represents about 1 lb of water per lb of lignite, or 30 gallons of water per million Btu in the product gas. The associated air separation plant would produce about 3,000 tons per day of oxygen. Such plants will produce pipeline gas plus by-products such as sulfur, light oil, ammonia, and ash. The highly aromatic light oil can easily be gasified if it is not desired as a by-product. Total recycling of water will be practiced in the plant, and emissions from the plant will be minimal.

HIGH TEMPERATURE WINKLER GASIFICATION

Introduction

In 1922, the patent concerning a fluidized-bed reactor for coal gasification was first applied for by F. Winkler. This unit showed marked advantages compared with the gasification processes known until that date. On the basis of the simple gasifier concept which can be easily operated, a substantial increase in capacity could be achieved entailing the erection of a number of large-scale industrial plants.

From 1956 to the end of 1964, two fluidized-bed generators with a capacity of 17,000 Nm^3/h (about 600,000 scf/h) each were operated by the Union Rheinische Braunkohlen Kraftstoff AG, a subsidiary of the Rheinische Braunkohlenwerke AG, for the production of synthesis gas. In these plants, Rhenish brown coal was gasified. The corresponding operating data are shown in Table 6.8.

Table 6.8: Operating Data at a Winkler Generator

Coal
Rhenish Brown Coal 0–8 mm (59% less than 1 mm)
Elementary Analysis of the Net Coal (maf)

Component	Weight Percent
C	68.1

(continued)

Table 6.8: (continued)

Component	Weight Percent
H	5.2
O	25.2
N	0.5
S	0.5
Water	8-10
Ash	4-5

Application
Related to 1,000 Nm^3 Product Gas

Coal with 8.5 wt % H_2O	645 kg
Oxygen (100%)	235 Nm^3
Steam for gasification	50 kg
Steam for cooling of the generator wall	50 kg
Condensate for gas	60-110 kg

Products
Gas Composition at a Production Rate of 17,000 Nm^3 gas/h

Component	
CO	40.2 vol %
CO_2	13.8 vol %
H_2	35.3 vol %
CH	1.8 vol %
N_2	0.9 vol %
Dust	40 kg/1,000 Nm^3 gas
Steam	485 kg/1,000 Nm^3 gas

Source: NTIS GFERC/IC-77/1

Process Description

Based on these experiences, the Rheinische Braunkohlenwerke AG decided to continue the development of this gasification technology. The aim of this development is to increase reaction pressure and temperature in the gas generator.

Coal gasification in a fluidized bed under pressure offers the possibility to provide the gas consumer, for example, an iron ore reduction plant (blast-furnace or direct reduction) or a power generating station, directly in one and the same heat with the product gas. Apart from these advantageously simple application possibilities of a gas generator working under pressure, an increase in the unit capacities of each gas generator and in cost savings relating to investment are expected.

Thermodynamic calculations have shown that elevated gasification temperatures (up to 1000°C) are required to reduce the content of oxidizing components, such as H_2O and CO_2, a fact which is important if such a gas is used in the metallurgical industry, for example. At the same time, the methane content is reduced and higher hydrocarbons are decomposed.

The use of high gasification temperatures involves the risk of ash sintering and sticking, with various negative effects on operation. Preliminary tests in the Institut für Eisenhüttenkunde (Institute for Metallurgy) of the RWTH Aachen (Technical University of Aachen) have shown that the addition of limestone will raise the softening and melting point of the ash substance, for example, of Rhenish brown coal, by forming a so-called granular slag. Further, the bonding of sulfur components by the lime agent is a useful effect caused by the limestone addition.

Process Development

The feasibility of this process, called the High Temperature Winkler (HTW) process and illustrated schematically in Figure 6.12, has been confirmed in its principles in an experimental plant of the RWTH Aachen (9).

Figure 6.12: Simplified Schematic of High Temperature Winkler Process

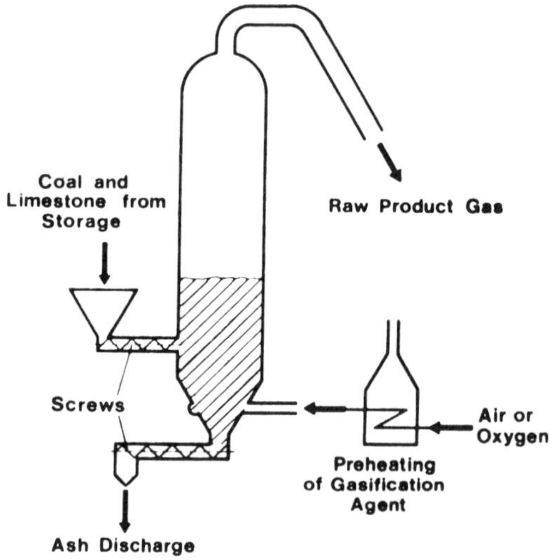

Source: NTIS GFERC/IC-77/1

Simultaneously, tests relating to the reaction kinetics of gasification were carried out, the results corresponding exactly with the data obtained from model calculations. The results of the experiments carried out at the RWTH Aachen are providing the basis for the planning of a test plant which is currently being erected by the Rheinische Braunkohlenwerke AG near Cologne. Characteristic design data of this HTW test plant are as follows:

Coal input	1,000 kg/h maximum
Gasification agents	O_2/air/steam
Gas generation rate	70 kmol/h maximum
Pressure	11 bars maximum
Gasification temperature	up to 1100°C
Starting-up	Autumn 1977

The HTW test plant (Figure 6.13) is supplied with dried brown coal containing about 18% of water which is stored and redried to reach a water content of 8 to 12%.

Figure 6.13: Flowsheet of the HTW Pilot Plant

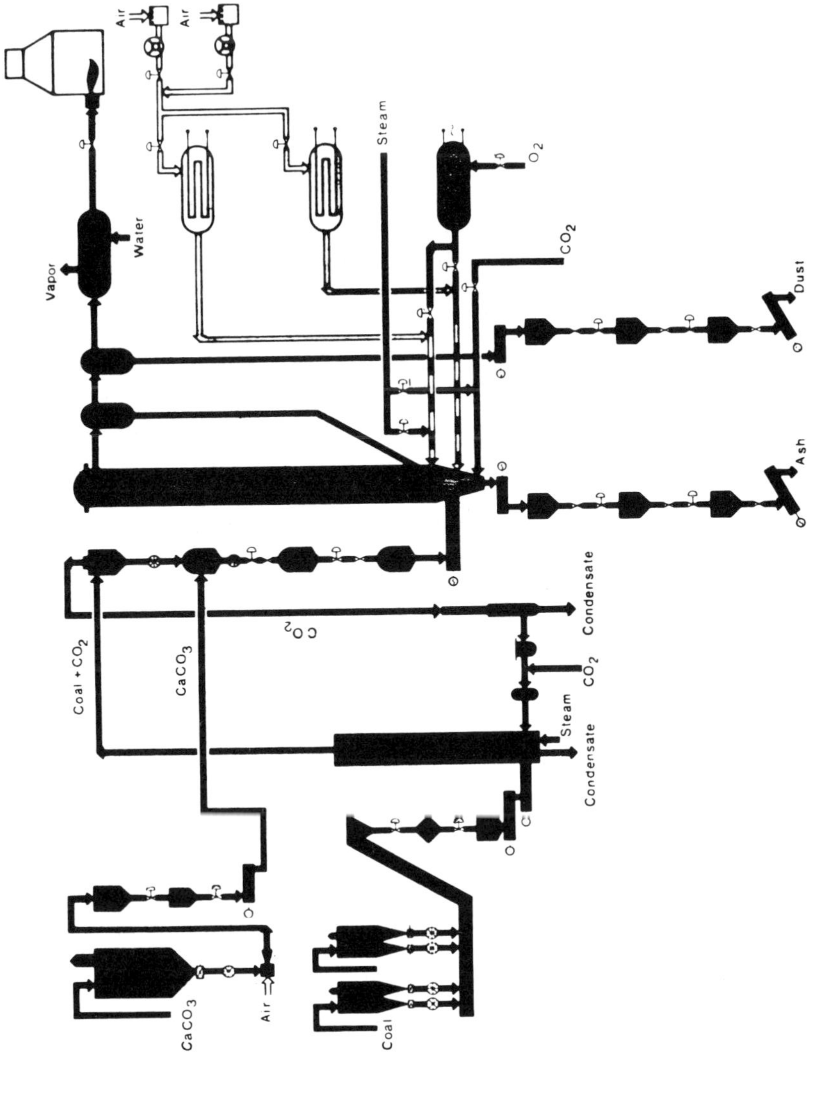

Source: NTIS GFERC/IC-77/1

In a lock-hopper system, the coal is brought to reaction pressure. At the same time, quantities of dolomitic limestone or lime are fed into the reactor by means of a screw conveyor. In the gasifier, the coal reacts with the preheated gasification agent, which is composed of air, oxygen, and steam.

The product gas leaves the gas generator at the top carrying away a part of solid particles which are separated afterwards in two cyclone precipitators. The first-separated solid particles are fed back to the fluidized bed in the reactor. Ash dust is removed in a second cyclone and discharged through a system of locks.

In case of normal use, for example, in the metallurgical industry for iron ore reduction by means of gas from brown coal, the product gas would be fed to the reduction plant after having passed the second cyclone. In this pilot plant, the purified product gas then flows through a waste heat boiler, where it is cooled to be passed on to an electric power plant. The gasification plant is designed so as to serve not only tests with respect to brown coal but also to other types of coal, such as hard coal.

Engineering as well as civil works and erection management of the HTW test plant have been taken over by Friedrich Uhde GmbH, Dortmund, West Germany. The project is subsidized by the Federal Ministry for Research and Technology of the Federal Republic of Germany.

Application of HTW Process

The HTW Process provides the possibility for the production of reduction gas with a low sulfur and dust content and an oxidation degree lower than 10%. On account of this gas quality and the pressure operation, the gas can be fed to a reduction process in one heat, i.e., without cooling or conditioning. After having been used for ore reduction, reduction gas, which is produced by means of air, is best suited for power generation in the blast furnace gas industry. When reduction gas is produced using oxygen instead of air, its use for ore reduction in a shaft furnace involves a regeneration cycle for the top gas from the shaft furnace.

In a procedure similar to the original Winkler process, it is possible to generate, under consideration of the advantages of the HTW Process, a type of gas suitable for use as raw synthesis gas for ammonia or methanol synthesis.

When using air as gasification agent, a low Btu gas can be generated. Using this low calorific gas under pressure in combination with a gas turbine, the thermal efficiency of a power station can be improved. The desulfurization of the gas to be achieved in the HTW Process results in a reduction of environmental pollution. In the field of power generation, this HTW-gas turbine process is an alternative procedure to flue gas desulfurization.

HYDROGASIFICATION OF BROWN COAL USING PROCESS HEAT FROM NUCLEAR REACTORS

Introduction

Hydrogasification of coal is part of the research and development activities of

Rheinische Braunkohlenwerke AG, Cologne. Rheinische Braunkohlenwerke is the most important brown coal producer in the Federal Republic of Germany, with an annual production of about 110 million metric tons (about 121 million short tons).

In the field of gasification, Rheinische Braunkohlenwerke is investigating, on one hand, processes in which the necessary heat for the gasification process is supplied by combustion of coal, so-called conventional gasification (see High Temperature Winkler Gasification), and, on the other hand, gasification processes in which process heat from high temperature nuclear reactors is used for gasification, so-called nuclear gasification.

Nuclear Gasification

In nuclear gasification, Rheinische Braunkohlenwerke is developing hydrogasification as part of the energy research program of the Ministry of Research and Technology of the Federal Republic of Germany. Within this program, steam gasification of coal using heat from high temperature nuclear reactors is developed by the Bergbau-Forschung GmbH, Essen. A general scheme of the hydrogasification process in connection with a high temperature reactor is shown in Figure 6.14.

Figure 6.14: Hydrogasification of Brown Coal

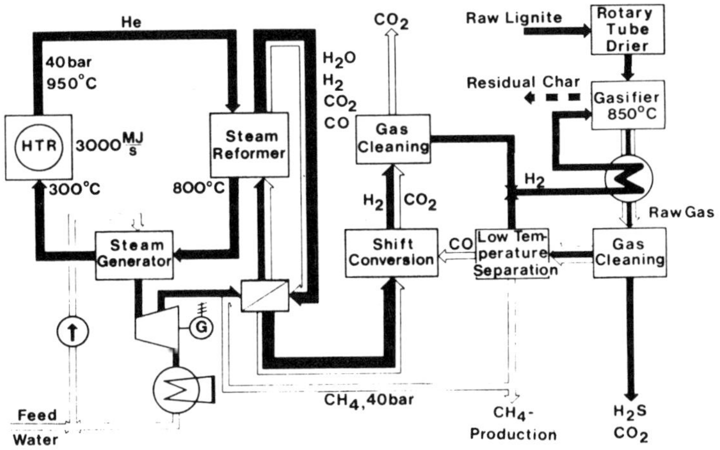

Source: NTIS GFERC/IC-77/1

Most of the heat required for the gasification process is coupled out from the nuclear loop by a steam reformer, heated by helium, in which methane with steam is reformed in an endothermic reaction into a synthesis gas consisting mainly of carbon monoxide and hydrogen. In this process step, on one hand, the supply of process heat for the gasification is achieved; on the other hand, the hydrogen balance of the gasification process is closed.

Raw brown coal is dried to a moisture content of 10% and then fed into the gasifier, where it is gasified with hydrogen in an exothermic reaction in a fluidized bed at 850°C (1562°F) and 80 bars (1,160 psi). Carbon dioxide and hydrogen sulfide in the raw gas are removed. A low temperature separation splits the cleaned raw gas into methane, hydrogen, and carbon monoxide fractions. The hydrogen is recycled to the gasification reactor as the gasifying agent. The larger part of the methane product is compressed to 70 bars (1,015 psi) and disposed of as synthetic natural gas (SNG). The smaller part of the methane is fed into the steam-reformer where it is reformed in an endothermic reaction into carbon monoxide and hydrogen at 800°C (1472°F). The required reaction heat is supplied by nuclear process heat. The reformed gas is fed, together with the carbon monoxide from the low temperature gas separation unit, into a shift-conversion unit and finally to the gasifier.

Hydrogasification by Rheinische Braunkohlenwerke AG

As far as gasification itself is concerned, Rheinische Braunkohlenwerke has been operating a hydrogasification plant with a capacity of 200 kg/h (440 lb/h) dried brown coal since May 1975. A simplified flow scheme of the pilot plant for hydrogasification is shown in Figure 6.15.

Figure 6.15: Pilot Plant for Hydrogasification of Coal

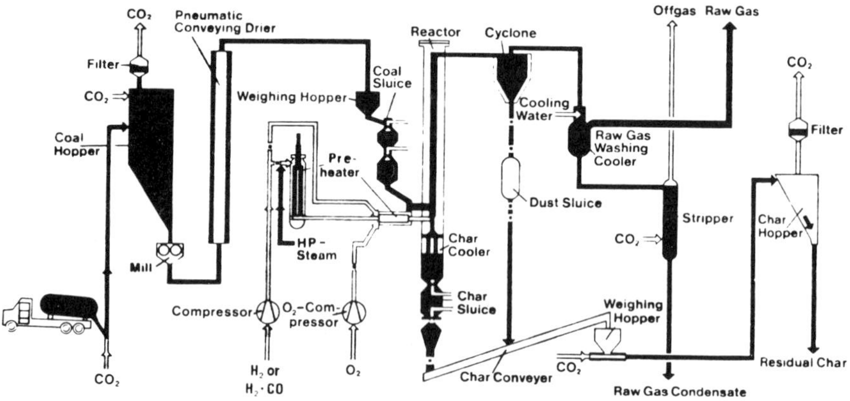

Source: NTIS GFERC/IC-77/1

The delivered coal is stored, ground, dried, and fed to the gasifier by a lock hopper system. The gasifier itself is a fluidized-bed reactor, designed for maximum pressure of 100 bars (1,450 psi) and temperatures up to 1000°C (1832°F). The gasifier has an inner diameter of 200 mm (7.9 in) and a height of 8 m (26.2 ft); the height of the fluidized-bed itself amounts to about 4 m (13.1 ft).

Hydrogasification Process

The reaction of coal with hydrogen is in two steps: in the first (fast) step, the volatile coal components are hydrogenated; in the second (slow) step, the fixed carbon is hydrogenated. For the total conversion of coal to methane, a rather long residence time of the solid material is required. For economic reasons, a total coal conversion is not envisaged; the finally received residual char is removed when the coal conversion of 70% is achieved. The residual char is withdrawn, cooled down, and depressurized in a lock hopper system which is similar to that in the feeding system. The residual char of the hydrogasification in general can be stored and used for further purposes, e.g., in combustion processes to generate electricity or for steam gasification processes.

The raw gas is mechanically cleaned by an internally insulated cyclone and cooled down by direct water scrubbing.

A specially designed hydrogen sulfide removal unit and a methane separation unit, which would be used in a commercial plant, are not installed in this pilot plant. The raw gas of the pilot plant is disposed of as a fuel gas.

An example of some important characteristic data during a test period in mid-1976 is shown in Figure 6.16. The gasification pressure during the test period was 80 bars (1,160 psi), the feed-charge of hydrogen was about 400 Nm^3/h (14,125 scf/h), and the coal feed rate approximately 160 kg/h (353 lb/h) of dried brown coal.

Figure 6.16: Test Run at Hydrogasification Pilot Plant

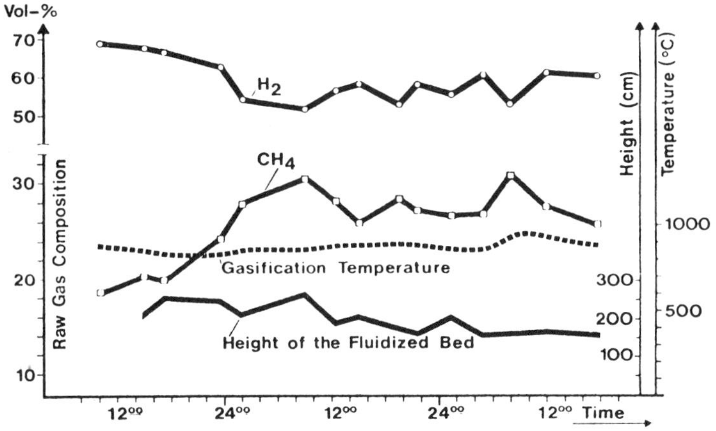

Source: NTIS GFERC/IC-77/1

At a fluidized-bed height of 150 to 250 cm (59 to 98 in) and a reaction temperature of 850 to 900°C, a methane content of the raw gas up to 30 volume % was obtained. Further results of improved tests in January 1977 are shown in Table 6.9.

The increased methane content of raw gas had been possible by decreasing the hydrogen velocity relative to feed rate and increasing the coal throughput up to 300 kg/h (dried brown coal).

Table 6.9: Gasification Test Run Data

	Conditions		
	1	2	3
Gasification temperature, °C	890	865	920
Gasification pressure, bar	80	80	80
Height of fluidized bed, m	2.7	2.7	2.6
Input of brown coal (mf), kg/hr	265	300	212
Grain size, mm	<1	<1	<1
Input of hydrogen, Nm^3/hr	435	384	382
	Results		
Carbon conversion, %	54.8	50.6	63.2
Composition of raw gas (dry, N_2 free), vol %:			
CH_4	31.2	34.2	34.3
C_{2+}	2.3	3.4	2.5
H_2	58.6	52.8	55.2
CO_2	1.7	3.1	1.9
CO	6.2	6.5	6.1
Yield of hydrocarbons, %	0.82	0.81	0.83
Residence time of char, min (~)	19	16	28

Source: NTIS GFERC/IC-77/1

Summary

Operational results obtained by the High Temperature Nuclear-Reactor AVR, of the Single-Tube Methane-Steam-Reformer Plant and the Hydrogasification Pilot Plant, have proved the feasibility of gasification of coal to SNG by means of nuclear process heat.

Compared with conventional gasification, advantages are expected concerning environmental aspects, coal consumption, and gas production costs. As far as the coal consumption is concerned, in conventional gasification a consumption of 6 to 7 kg (13.2 to 15.4 lb) of raw brown coal (60% H_2O) per Nm^3 (35.3 scf) of CH_4 is to be compared with a consumption of 4 kg in a gasification process based on nuclear process heat, which means a coal saving of about 40%.

REFERENCES

(1) Scott, J., "Westfield Development Centre—Concept and Facilities," *Proceedings of Sixth Synthetic Pipeline Gas Symposium*, American Gas Association, Chicago, IL, October 28-30, 1974.

(2) Hebden, D., "High Pressure Gasification Under Slagging Conditions," *Proceedings of Seventh Synthetic Pipeline Gas Symposium*, American Gas Association, Chicago, IL, October 27-29, 1975.

(3) Gronhovd, G.H., Harak, A.E., Kube, W.R., and Oppelt, W.H., *Design and Initial Operation of a Slagging, Fixed-Bed, Pressure Gasification Pilot Plant*, BuMines, RI, 6085, (1962).

(4) Gronhovd, G.H., Harak, A.E., Fegley, M.M., and Severson, D.E., *Slagging Fixed-Bed Gasification of North Dakota Lignite at Pressures to 400 psig*, BuMines, RI, 7408 (1970).

(5) Schora, F.C., Lee, B.S., and Huebler, J., "The Hygas Process," paper presented at the 12th World Gas Conference and Exhibition, Nice, France, June 5-9, 1973.
(6) Schora, F.C., Lee, B.S., and Tobin, D.J., "The 'New' Coal Gasification Technology," paper presented at the Conference on Synthetic Hydrocarbons, Annual AIME Meeting, Dallas, February 24-27, 1974.
(7) Lee, B.S. and Lau, F.S., "Results from Hygas Development," paper presented at the 77th National Meeting of the AIChE, Pittsburgh, June 2-5, 1974.
(8) Bair, W.G., Leppin, D., and Lee, A.L., "Design and Operation of Catalytic Methanation in the Hygas Pilot Plant," paper presented at the American Chemical Society, Division of Fuel Chemistry Meeting, Atlantic City, N.J., September 9, 1974.
(9) Flesh, W. and Velling, G., "Coal Gasification in the Winkler Gas Generator," *Erdol und Kohle, Erdgas, Petrochemie* 15, pp. 710-713 (1962).

LIGNITE LIQUEFACTION

The material in this chapter has been based on:

Technology and Use of Lignite, Proceedings of a Symposium Sponsored by the U.S. Energy Research and Development Administration and the University of North Dakota, Grand Forks, ND, May 18-19, 1977 (NTIS GFERC/IC-77/1), compiled by G.H. Gronhovd and W.R. Kube, Cochairmen, for the U.S. Department of Energy, 1977

Flash Hydrolysis Process for Conversion of Lignite to Liquid and Gaseous Products (ERDA Report CONF-760401-5), prepared by M. Steinberg, T.V. Sheehan and Q. Lee of Brookhaven National Laboratory for the U.S. Energy Research and Development Administration, October 1975

"The Economics of Carbonizing Texas Lignite," *Energy Communications* 5(5), 303-330 (1979), by M.E. Leesley of the University of Texas Department of Chemical Engineering and N.A. Massey of AAA Technology, Houston, TX

"Carbonizing Texas Lignite: The Technical Feasibility," by M.E. Leesley of the University of Texas Department of Chemical Engineering and E.F. Aul of Conoco Development Company (unpublished).

CO-STEAM PROCESS

Introduction

CO-Steam is a second generation coal liquefaction technology for converting low rank coals to environmentally acceptable liquid boiler fuel, and with fractionation or additional hydrogenation, to distillate heating oil, gasoline, and chemical feedstocks. The process utilizes the reaction of carbon monoxide (or synthesis gas) and steam with or without the addition of catalyst. Favorable reaction conditions are 4,000 psig and 460°C. Work on the CO-Steam process at the Grand Forks Energy Research Center (GFERC) was started in 1975.

The conversion of coal to oil by reaction with carbon monoxide and steam was first reported in 1921 by Fischer and Schrader (1), who believed that the liquefaction occurred through the reaction of hydrogen generated in situ by the water-gas shift reaction. This work was extended in batch autoclave studies by Appell (2)(3) at the Pittsburgh Energy Research Center (PERC), who reported with regard to reaction mechanism:

(1) that alkali formates are intermediates in the shift reaction;
(2) that the sodium and calcium in lignite catalyze the reaction;
(3) that utilization of hydrogen depends on the use of a suitable solvent; and
(4) that CO reacts to remove carbonyl oxygen from lignite.

Appell also reported that an increase in temperature between 450° and 475°C promoted cracking and reduced product viscosity, with a slight attending increase in utilization of hydrogen (4).

Three facets of the previous work are shown to be particularly important by the recent batch autoclave studies at GFERC: (1) the removal of coal oxygen by CO, (2) the action of the hydrogen-donor solvent, and (3) the effect of temperature above 460°C.

The CO-Steam reaction has been shown to be promoted by addition of sodium formate, alkali carbonates, iron oxide, metal chlorides, or nitrogen bases in work performed at PERC (3)(5) and at Montana State University (6)-(9). Aspects of this work are being continued at PERC under the project on throw-away catalysts and at GFERC under the investigation on the effect of the naturally occurring mineral content in lignite.

The development of a continuous CO-Steam process was initiated by Yavorsky and Del Bel at PERC (10), and is being continued at GFERC. The first studies at PERC were conducted in a 1-liter stirred reactor at a throughput of about 1 lb coal/hr in anthracene oil; benzene-soluble oil yields of about 60 wt % of maf coal were obtained using synthesis gas at 4,000 psig, 425°C, and 2-hour residence time. More recent experiments at PERC (1976), using a tubular reactor with a throughput of about 5 lb coal/hr, have resulted in a lined-out product that is less viscous but still heavier than desired.

Design and construction of a 3- to 5-lb coal/hr continuous CO-Steam process unit at GFERC was started in 1975, and checkout and shakedown operation are currently in progress. This facility has been designed to provide flexibility for changing backflow (residence time distribution) and level of turbulence to permit the systematic study of improvements in reactor design. Supporting studies are already underway at GFERC involving the determination of reaction kinetics in a time-sampled batch autoclave and extensive work on product characterization using an array of dedicated analytical instrumentation.

Future development of the CO-Steam process at GFERC will be principally concerned with reactor design and the attainment of desired product characteristics, which include high yield, low viscosity, and reduced contents of sulfur, nitrogen, and ash. The primary product should be suitable for pipeline transport and for burning in power plants without stack gas cleaning. Process economics are believed to be favorable compared with processes involving catalytic hydrogenation of bituminous coals owing to lower costs for lignite and synthesis gas. Preliminary results in batch experiments at GFERC, presented in this paragraph, offer encouraging prospects for a higher-temperature, short residence time process, which would further improve economics. If these findings can be confirmed in the continuous process unit, early scale-up in a 1- to 10-ton/day Process Development Unit (PDU) would be warranted.

Studies on Batch Reaction Kinetics

The purpose of these tests is to obtain rate data showing the effect of key operating variables in relation to time, hopefully to find conditions which produce a high yield of liquid product in minutes rather than in an hour or longer. Previous batch experiments on CO-Steam had been performed with an extended warmup after charging reactants, which rendered impractical the study of the rapid reactions that might be occurring.

Rates are determined for reactions of CO and H_2 with slurries of lignite and solvent oil in the presence of water. Changes in both the gas and the slurry are measured as a function of time, starting with the first minutes of reaction.

The main variables studied to date at GFERC have been the reactant gas composition, the solvent, and the temperature. Pressure has been varied between 3,000 and 4,800 psig. Changes in gas composition during reaction are determined by on-line gas chromatographs; liquid products are characterized by extraction, distillation, and analysis of fractions using liquid chromatography, mass spectrometry, and other selected methods.

Time-Sampled Autoclave

Studies on batch kinetics are carried out in a 1-liter magnetically stirred autoclave, which is specially outfitted for rapid charging of slurry to the preheated reactor and for timed sampling of gas and slurry during the course of an experiment. Hot charging brings the slurry to reaction temperature within 3 minutes, thereby eliminating the 1- to 2-hour warmup; alternately, the slurry can be charged at an intermediate temperature and heated to observe the effects of temperature change. Sampling of gas and liquid can be started 5 minutes after charging and continued at nominally 10-minute intervals to follow the course of the reaction.

The important features of the reactor and on-line gas chromatograph are shown in Figure 7.1. Reactant gases (CO or H_2) or purge gases (N_2 or He) are charged from cylinders, either directly to the reactor or via a compressor and accumulators. Gas is also used to displace the coal-oil slurry from the charging cylinder into the reactor; 200 to 400 g of slurry containing 30 to 48 w/w % as-received lignite (100% -60 mesh) are charged to the hot reactor. Reactor contents are stirred by a turbine mixer turned at 500 rpm by a magnetic drive. Samples of the reacted slurry (about 1 ml) are drawn from the bottom of the reactor through a pneumatically operated double valve system.

Gas samples are taken from the side of the reactor and introduced into a heated sampling system for analysis in three on-line gas chromatographs (Figure 7.1). Samples are initially drawn into an evacuated volume of about 1 ml, maintained at 400°C in order to retain the contents of volatile organics, water vapor, and gas that are representative of the reactor temperature and pressure.

Figure 7.1: Schematic for the Time-Sampled Batch Autoclave and On-Line Gas Chromatograph System

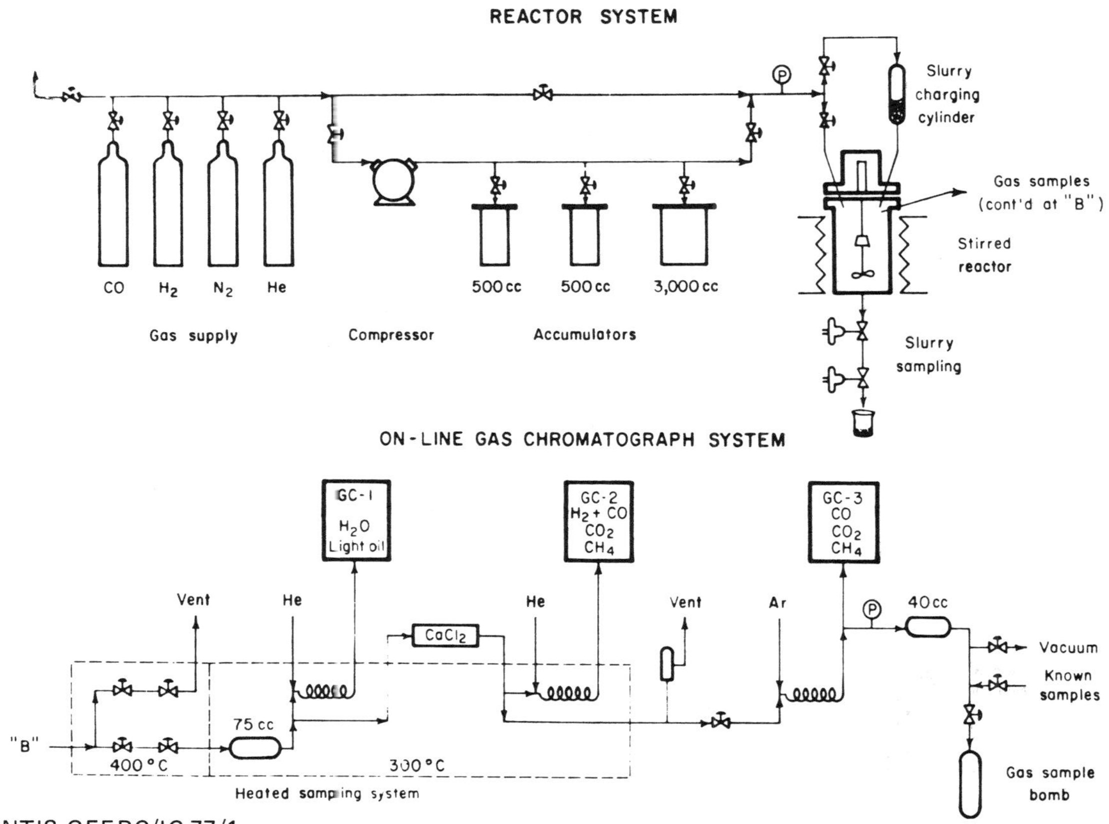

Source: NTIS GFERC/IC-77/1

The sample is then expanded to atmospheric pressure and cooled first to 300°C and then to room temperature. Helium carrier gas is used to transfer a known volume of the reactant from a 300°C sample loop to the first gas chromatograph (GC) to analyze for water and light organic liquids. The remainder of the sample is filtered through $CaCl_2$ at room temperature to remove the water and organic liquids before analyzing for H_2, CO, CO_2, CH_4, C_2H_6, and C_3H_8 in two additional GCs. Sample gas can also be collected in a bomb for later analysis by mass spectrometry.

The charging and sampling systems briefly described here were developed at GFERC after extensive trial operation to find components and conditions that would function reliably. The system has been continuously improved, and sampling can now be accomplished entirely remotely by time-sequenced servos controlled from the panel board outside the barricade.

Analytical Methods

Reactor slurry samples or fractions thereof are analyzed to determine: (1) the major product fractions (gas, oil, and solid residue), (2) molecular weight distribution, (3) compound types, (4) molecular composition, and (5) selected elements. The sample and analysis tree shown in Figure 7.2 depicts the principal methods used and the types of information obtained. Similar analyses are being performed at GFERC on a number of coal liquefaction products from other projects to gain insights into the broad relationships between product quality and processing conditions. Reports have been published or are in progress both on methods (11)(12) and on the analytical data obtained (13)(14).

The major product fractions, which are discussed later, are defined by the separations used for their analysis. Gas yields are determined from reactor pressures, the gas analyses from the on-line GCs, and STP volume measurements. The solid residue is determined by extraction at room temperature with tetrahydrofuran (THF) and filtration through a 0.5-micron filter. The nondistillable fraction, including a portion of the soluble oil, is determined by microdistillation performed on the slurry at 250°C and 1 torr. The oil distilled from the slurry includes material derived both from coal and from the starting solvent; that derived from coal is determined by difference after suitable corrections for production of gas.

The analyses performed on various fractions of the liquid product or residue (Figure 7.2) at present provide indices of product quality.

One of the most important indices of quality has been the distribution of molecular weight for the heavy (nondistillable) fraction of the oil product. The molecular weight is determined by high pressure liquid chromatography (HPLC) using a Water's ALC-GPC 201 system, fitted with micron styragel columns and calibrated with polypropylene glycol standards. The technique leads to compression of material above molecular weight 800, but in the range of 200 to 800, it is accurate within 5% of values measured by mass spectrometry or vapor phase osmometry.

Analysis of the molecular composition of CO-Steam product is accomplished using mass spectrometry as the primary tool. Both identification of the compounds present and the measurement of their concentrations are important in evaluating product quality, reaction rate and mechanism, and potential health hazards.

Figure 7.2: Analysis Schematic for Autoclave Slurry Samples and Other Coal Liquefaction Products

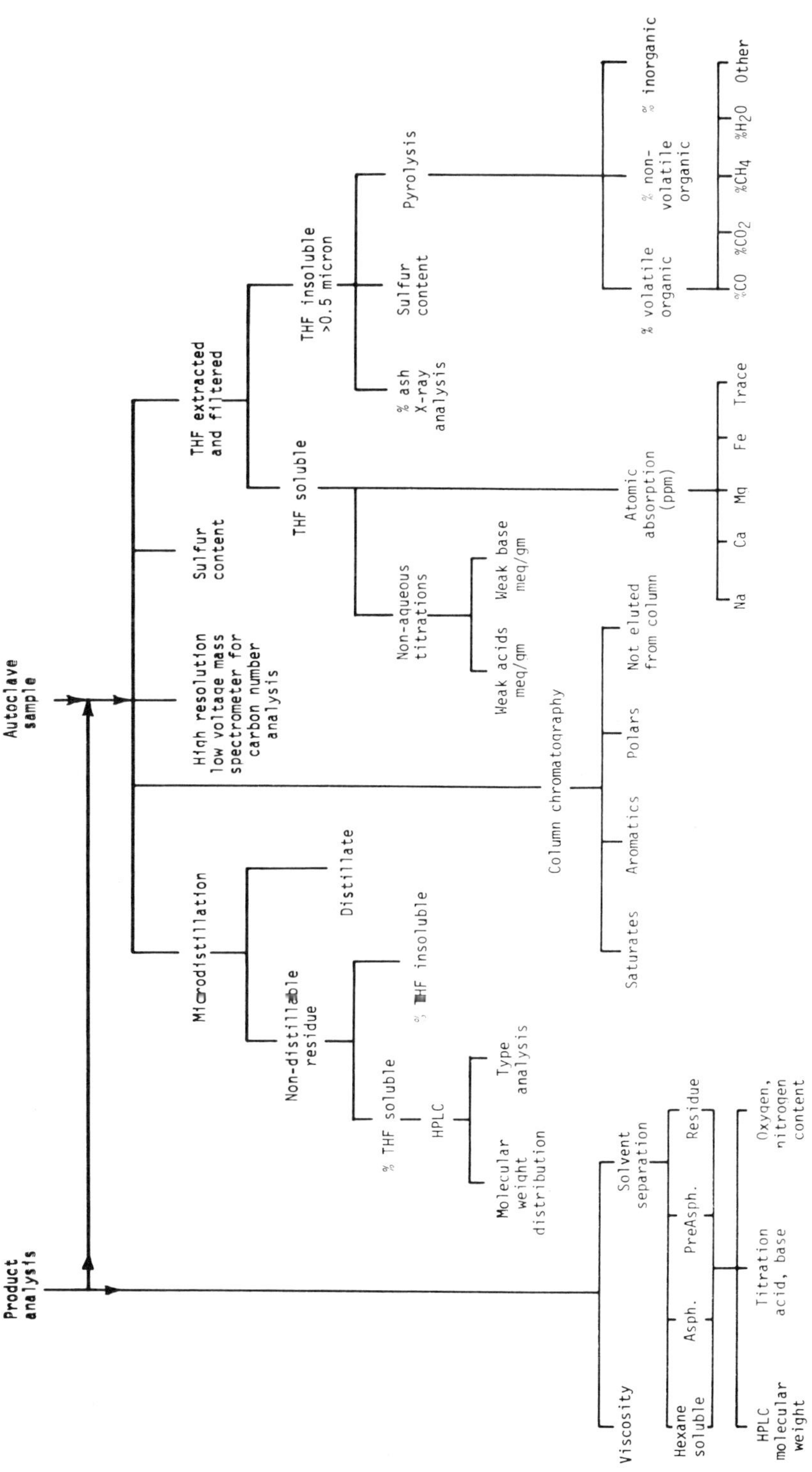

Source: NTIS GFERC/IC-77/1

These techniques give compositional information for compounds up to molecular weight 378, and molecular weight data for species up to 700 to 900 molecular weight.

Compounds are identified by a combination of column chromatography with gas chromatography-mass spectrometry (GC-MS) and/or high resolution mass spectrometry (HRMS). The whole sample is fractionated into saturated hydrocarbons, aromatic hydrocarbons and ethers, nitrogen compounds, and hydroxyl compounds using an alumina column (neutral grade, activity 1) and sequential elution by hexane, toluene, chloroform and tetrahydrofuran/ethanol. The isolated fractions are simpler mixtures of specific chemical types, and when subjected to GC-MS, absolute identification of most major compounds is achieved. Components in mixtures which are too nonvolatile for GC-MS analysis are identified by HRMS.

Quantitative analyses are generally accomplished on unseparated samples by low voltage mass spectrometry (LVMS). Using low ionizing voltage causes only molecular ions (parent ions) to appear in the spectrum. The output sensitivities for compounds known to be in the CO-Steam product have been measured for our instrument, so the concentration of each compound observed is calculated using the sensitivity and ion intensity. Results are normalized and corrected for saturated hydrocarbons and nonvolatile content, since only unsaturated volatile compounds are observed. Computerized data collection and handling allow the entire analysis, calculation, and tabulation of data to be done in less than one hour.

Other important techniques include: (1) nonaqueous titration for weak acids and bases, (2) type analysis by HPLC, (3) UV and IR spectroscopy, (4) mineral analysis by atomic absorption, (5) pyrolysis of solid residues followed by GC analysis on the volatiles, and (6) microelemental analysis. The role of elemental analysis has been limited due to problems encountered in finding equipment and techniques which provide reliable microanalysis on nonvolatile samples; development is continuing because of the importance of having a rapid method for obtaining elemental material balances in future experiments on the continuous processing unit.

Extent of Rate Control by Mixing in the Batch Reactor

In a multiphase system such as CO-Steam, it is possible that the overall rate of reaction between gaseous reductants and the coal in the slurry could be controlled by one or more of the diffusional steps that bring reacting gas to the surface of the coal or products back to the gas phase, rather than by chemical reaction. In order to determine whether mixing in the batch reactor was sufficient to prevent control by diffusion between the gas and the liquid, analyses were performed on the gas dissolved in the slurry liquid. If these analyses were similar to the analyses in the bulk gas, then the controlling step could be judged to be either chemical reaction or liquid diffusion at the coal surface.

Aliquots of slurry were collected in a 1.77 ml sample bomb and the dissolved gas expanded into the GC system to analyze for H_2, CO, CO_2, and CH_4. Ratios of gas analyses (dissolved gas versus bulk gas) are given in Table 7.1 for an extended experiment involving step changes in temperature from 400° to 435°, to 475°, to 400°C. The gas charged was 47.5 mol % H_2 and 53.3 mol % CO; approximately 70% of the CO and 14% of the H_2 were consumed during the experiment.

Table 7.1: Gas Solubility and Ratios of Analyses for Dissolved Versus Bulk Gas in the Batch Autoclave*

Time, (min)	Temperature, (°C)	Pressure, (psig)	Gas Solubility, [ml (STP)/g Slurry]	Ratios of Gas Analysis: . . Dissolved Gas/Bulk Gas. . . H_2	CO	CO_2	CH_4
92-113	400	4580	35	0.88	0.73	1.81	2.30
141-178	435	4490	37	0.87	0.74	1.38	1.10
200-248	475	4240	35	0.85	0.64	1.15	0.96
285-309	400	3620	37	0.80	1.12	1.38	1.45

*Run 30

Source: NTIS GFERC/IC-77/1

The amount of gas dissolved in the slurry was not affected by time or temperature and remained essentially constant at 35 to 37 ml (STP) per gram of slurry. However, ratios of analyses in general showed depletion of reactant gases (CO and H_2) and enrichment of product gases (CO_2 and CH_4) during the reaction periods at 400°, 435°, and 475°C. Reaction was essentially complete by the time the reactor had cooled to 400°C, and these ratios should therefore tend to reflect the equilibrium solubilities at this temperature.

Effects of Solvent on Conversion and Coking

Experience with charging and operating the batch autoclave at high temperatures (up to 480°C) has demonstrated the need for a solvent that remains substantially in the liquid phase under conditions existing in the reactor and that has good hydrogen donor characteristics. At 440°C or above, use of anthracene oil, which lacks hydrogen donor properties, has caused sampling lines to plug at the first attempt to take slurry from the autoclave, and reactor walls and stirrer have become coated with coke.

Use of Tetralin, which has good hydrogen characteristics, has resulted in segregation within the reactor due to vaporization of solvent and settling out of lignite. However, use of 10% Tetralin and 90% anthracene oil has yielded good results; sampling lines have remained open and no coking has occurred even after several hours of operation at 475°C with repeated charging of lignite. Segregation has not posed serious problems when using the 90:10 solvent, provided that the pressure in the reactor is maintained sufficiently high (4,000 psig at 475°C) and suitable proportions of lignite, solvent, and gas are charged.

Analyses of slurry samples taken from the batch autoclave indicate that 80% of the Tetralin charged in a 90:10 solvent had reacted by the time the first sample was obtained (7 minutes). Thereafter, the concentration of Tetralin remained nearly constant. This result indicates that a hydrogen donor such as Tetralin initially donates hydrogen rapidly, thereby promoting rapid liquefaction. The nearly constant level of Tetralin thereafter is believed to represent an equilibrium between the hydrogenation of naphthalene to Tetralin and the subsequent release of hydrogen from the Tetralin to the coal; this accounts for the observed action of Tetralin in continuing to promote conversion to both the distillable and the THF-soluble product categories (see Figure 7.3).

Figure 7.3: Conversion of Coal to Distillable Product and THF-Soluble Product

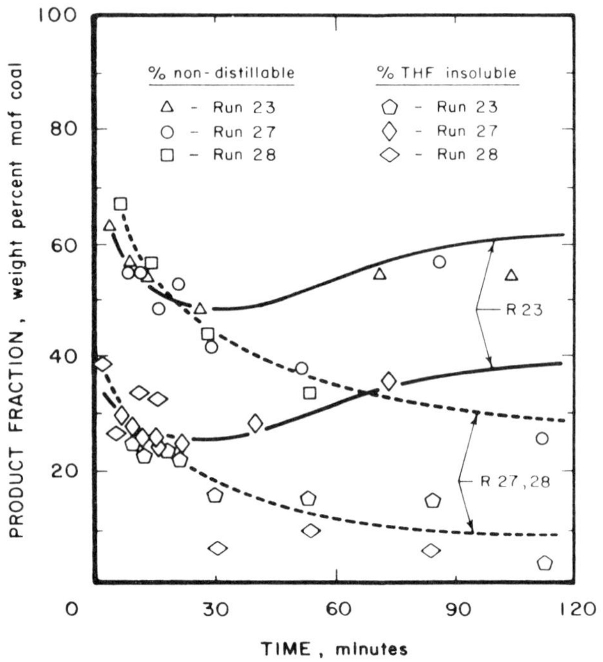

Note: Runs 27 and 28—10% Tetralin in anthracene oil as solvent. Run 23—Anthracene oil alone as solvent

Source: NTIS GFERC/IC-77/1

Presence of Tetralin also affected the character of the liquid produced with regard to the degree of alkylation of the aromatics present. Lined-out CO-Steam product from the continuous process unit at PERC exhibited a relatively low concentration of alkyl aromatics, whereas product produced using pure Tetralin as solvent in the batch unit at GFERC resembled a liquefaction product from catalyzed hydrogenation (Synthoil). The effect of a good hydrogen donor solvent in CO-Steam is thus shown to be similar to the effect of catalytic hydrogenation.

Rates of Reaction for Carbon Monoxide and Hydrogen

Experiments were conducted reacting CO, H_2, and 1:1 CO:H_2 synthesis gas with lignite over the range of temperature from 350° to 470°C. In addition, CO was reacted with water in the absence of lignite, and with dried lignite in solvent without water present. Reaction pressure was normally 4,000 psig.

Reactions of Carbon Monoxide: Representative results from experiments involving reactions of CO are shown in Figures 7.4 and 7.5, where reacted CO is plotted versus time.

Figure 7.4: Consumption of CO Versus Time at 440°C

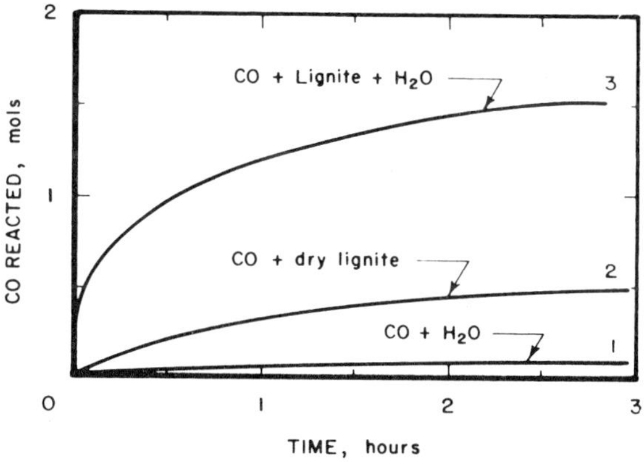

Source: NTIS GFERC/IC-77/1

Figure 7.5: Consumption of CO Versus Time During Heating from 350° to 470°C and Reacting at 470°C

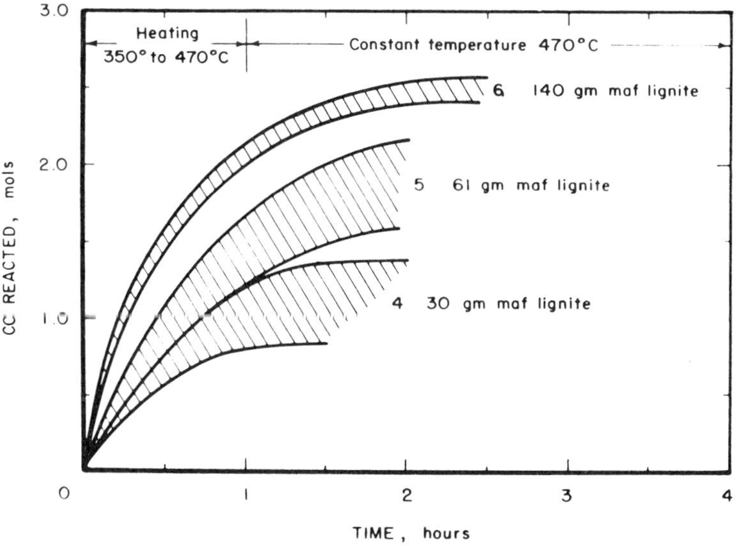

Source: NTIS GFERC/IC-77/1

In a starting mixture of CO and water in the absence of lignite (Figure 7.4, curve 1), the CO and water reacted very slowly to produce H_2. The CO remaining after 4 hours was 84% of that charged, compared with 42% that would remain at equilibrium at 440°C. A stepwise increase in temperature from 400°, to 440°, to 480°C (not shown) had no detectable effect on the rate of reaction.

This result clearly indicates that the production of hydrogen by reaction of CO and H_2O in the gas phase or on the surface of the reaction vessel is far too slow to be an important step in the liquefaction process at any temperature.

Reactions of CO in the presence of lignite (Figures 7.4 and 7.5) are initially very rapid during a period of about 30 minutes and thereafter tend to level off. The consumption of CO by reaction with dried lignite at 440°C (Figure 7.4, curve 2) is approximately one-third of that consumed in the presence of lignite and water (Figure 7.4, curve 3). The CO consumed by lignite in the absence of water (lignite was dried in solvent oil under nitrogen) is believed to represent removal of oxygen from coal, possibly carbonyl oxygen as reported by Appel (3). The difference in CO consumed between curves 2 and 3 in Figure 7.4 represents the amount of "nascent hydrogen" produced and reacted with the coal or solvent.

The amount of CO consumed in experiments shown in Figure 7.5 depends on the amount of lignite charged. For the smaller amounts of lignite, an appreciable portion of the CO remains unreacted (curves 4 and 5); whereas for the larger amount of lignite, essentially all of the CO is consumed (curve 6). The shaded regions in Figure 7.5 represent CO reacted to form hydrogen gas; the lower curve represents CO reacted to form "nascent hydrogen" or that reacting directly with coal. All of these curves involved one hour of heating to reach 350°C (zero time on the graph) with continued heating to 470°C in approximately one hour.

It is important to observe that the reaction of CO and lignite (lower curves) is the predominant reaction for the initial time period (zero to 30 minutes), which corresponds to the lower temperatures. The production of hydrogen gas via the water gas shift reaction is increased at the higher temperatures; however, the amount of hydrogen produced in curve 6 was limited by the small amount of CO remaining after reaction with coal (via nascent hydrogen or directly). This result demonstrates that CO can be caused to react predominantly with the coal (rather than being "wasted" by formation of hydrogen gas), provided that suitable ratios of lignite and CO are reacted at the right temperature.

Reactions of Hydrogen: Starting with 100% H_2 gas, the extent of reaction with lignite increases with temperature, and typically 50% of 4 mols of H_2 charged are reacted with 83 g of lignite in 2 hours at 440°C (Figure 7.6). The average mol rate of consumption of H_2 for the first 30 minutes of reaction is about the same as the rate of CO consumption under similar conditions.

In a mixture of 1:1 CO:H_2 (Figure 7.6, curve 2) far less H_2 is consumed in the early reaction owing to apparent preferential consumption of CO. The net rate of H_2 consumption from the gas is increased along with temperature, from essentially zero at 400°C to an appreciable rate at 470°C. Somewhat more H_2 than that shown may be reacting owing to formation of gaseous H_2 from CO and H_2O. The increased consumption at higher temperatures is believed to satisfy the hydrogen demand of free radical fragments resulting from thermal cracking, as discussed later in connection with data on reduction in molecular weight.

Starting with CO and H_2O in the absence of hydrogen (Figure 7.6, curves 3, 4 and 5) there is a net production of H_2 that depends on the presence of lignite and on temperature. There is a definite catalytic effect by lignite for that part of the water-gas shift reaction which acts to produce hydrogen in the gas phase.

Figure 7.6: Reactions of H_2 Versus Time

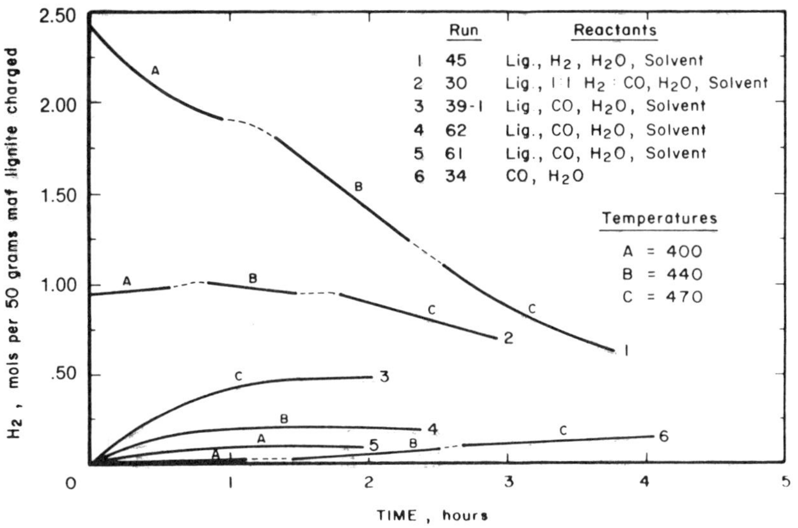

Source: NTIS GFERC/IC-77/1

Conversion of Coal to Product Fractions

Results of measurements on yields of major product fractions depend strongly on the analytical separation procedures that are used. Most previous work has relied on benzene extraction to determine soluble oil yield. The methods employed in the present study rely on both extraction and distillation. Whatever methods are chosen, it is important to obtain supplementary analysis on the quality of product fractions. Both yield and quality are discussed in the following sections.

The major product fractions (Figure 7.7) derived from the coal are: (A) the gas, (B) distillable oil product, (C) nondistillable but soluble oil product, and (D) solid residue. Gas yield (A) is determined from reactor pressure, with correction for volatilization of solvent, on-line GC analysis, and volume measurement. The insoluble residue (D) is determined by extraction at room temperature with tetrahydrofuran (THF) followed by filtration through a 0.5-micron filter. The nondistillable fraction is determined by distillation of an aliquot of total product at 1 torr and 250°C. The distillable oil product cannot be determined directly, since the volatile fraction includes material derived both from coal and from the starting solvent; the amount shown in region B in Figure 7.7 is, therefore, obtained by difference.

The conversion results shown in Figure 7.7 were obtained for reaction of lignite with 1:1 H_2:CO synthesis gas hot charged at 437°C and nominally 4,000 psig. The conversion of coal to gas and soluble product (A + B + C) is very rapid and reaches 70% in 10 minutes, followed by a slow increase to 90% after about two hours. The gas yield comprised of the coal oxygen appearing as CO_2, along with smaller amounts of methane, ethane, and propane, approaches 20% in 30 minutes and increases thereafter to about 30%. The yield of soluble oil (B + C) reaches 49% in 10 minutes and thereafter increases slowly to about 60%; the nondistillable portion (C) determines by its quantity and quality whether the total soluble product is a liquid or a solid. Higher reaction temperatures tend to reduce both the quantity of the "C" fraction and its average molecular weight, with an attending increase in distillable oil (B) and gas (A). In this experiment at 437°C, fraction C had an average molecular weight of about 500, which would cause the total soluble oil product (B + C) to resemble heavy pitch.

Figure 7.7: Conversion of Coal to Major Product Fractions in the CO-Steam Process

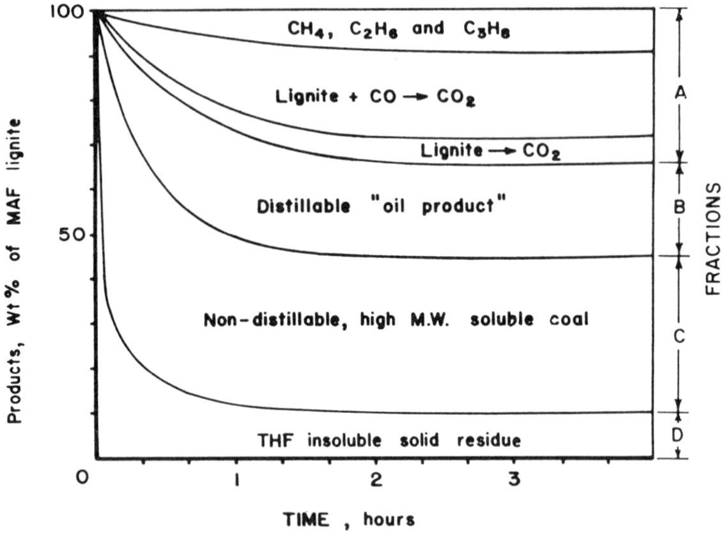

Source: NTIS GFERC/IC-77/1

Effect of Reaction Temperature on Molecular Weight Distribution

The molecular weight of the soluble but nondistillable product fraction (fraction C in Figure 7.7) has been observed to depend principally on temperature, and to a lesser extent, on reaction time. Characteristic changes in molecular weight distribution for stepwise increases in temperature from 400°, to 435° and to 475°C occur; the predominant peak in molecular weight is shifted from 1,170, to 410, to 280 molecular weight, respectively. Comparison of these molecular weight distributions with those for the distillation residues from other materials indicates that:

(1) product at 400°C resembles solvent refined coal,
(2) product at 435°C resembles lined-out CO-Steam product from the PERC continuous units operated at 425°C, and
(3) product at 475°C resembles SRC-II.

Products once produced at 400° and 475°C have been shown in other experiments to change only slightly with residence time up to 3 hours. At 435°C the product distribution does change with time, as evidenced by the slow disappearance of the 1,170 molecular weight peak over a period of 6 hours.

Product Viscosity

The viscosity of product oil correlates well with the quantity and quality of the nondistillable fraction. This has been demonstrated in preliminary studies at GFERC conducted on simulated oils prepared by dissolving distillation residues (250°C and 1 torr) from a number of coal liquefaction products in anthracene oil (15); products represented were solvent refined coals from the Pittsburg and Midway Coal Company, Southern Services, Inc., and Project Lignite at the University of North Dakota, and coal liquids from the CO-Steam and Synthoil processes at PERC. Since the anthracene oil used was nearly identical to the distillable fraction of the CO-Steam product, the prepared formulations closely approximated a series of products having differing heavy ends.

CO-Steam products from the continuous processing units at PERC operated at 425°C contained 40 to 45% nondistillable; product from a batch recycle study at GFERC at 475°C contained about 40%, indicating that this percentage is difficult to reduce by increasing temperature of the CO-Steam process. However, molecular weight is reduced along with an increase in temperature.

Effect of Temperature on the Mineral Content of Liquid Products

An interesting insight into the production of low molecular weight (280 MW) fragments from dissolved macromolecules is provided by the effect of temperature on the mineral content of the liquid product. The concentration of sodium (Na) has been rapidly reduced during the first 30 minutes of reaction at 400° or 437°C. However, upon further heating to 475°C, the concentration of sodium is first increased and subsequently decreased again with time. The temporary increase in sodium content occurs in coincidence with the shift to lower molecular weight discussed previously.

Since sodium is known to be dispersed uniformly through the organic matter of lignite, presumably as high molecular weight humic acid salts, the solubilization of sodium appears to reflect the depolymerization of these high molecular weight materials. The subsequent drop in concentration indicates that sodium later agglomerates and is rendered insoluble in the liquid product.

Characteristics of Distillate

The major compound types found in the material that can be volatilized in the mass spectrometer heated glass inlet at 350°C are shown in Table 7.2 for

(1) anthracene oil,
(2) CO-Steam batch product obtained at GFERC at 470°C,
(3) CO-Steam continuous product from PERC, and
(4) Synthoil from PERC.

The analyses were performed by low-voltage mass spectrometry which gives a carbon number analysis for each of the compound types indicated; values reported represent the group totals over all carbon numbers. The compounds represented in this analysis had been previously identified by column chromatography with gas chromatography-mass spectrometry or by high resolution mass spectrometry.

The CO-Steam batch product from GFERC was obtained by charging a slurry containing 48% lignite in a solvent comprised of 90% anthracene oil and 10%

Tetralin and reacting with 1:1 $CO:H_2$ at 470°C. In most respects, the analysis of the GFERC product falls between that of anthracene oil and of CO-Steam product from PERC, reflecting the fact that the GFERC product contains both coal-derived liquid and starting solvent.

The Tetralin charged in the solvent is shown to be almost completely converted to naphthalene, which accounts for the elevated value for the latter. Phenols are essentially absent in anthracene oil, and the quantity observed is a rough measurement of the extent to which the product is coal derived.

The amounts of other compound types are similar for the GFERC CO-Steam product and anthracene oil, owing to similarity between the coal-derived product and the anthracene oil; this makes precise determination of the kinetics of volatile product formation very difficult. The similarity does, however, indicate that anthracene oil is a good choice of solvent for batch studies on kinetics from the standpoint of obtaining results that are similar to those that would be obtained using a lined-out, coal-derived solvent.

The alternative of using a distinctly different solvent, such as Tetralin, would make the identification of coal-derived materials far easier but would also cause the liquefaction reaction to be nontypical.

Use of the mass spectrometer for studies on liquefaction kinetics would be greatly aided if materials of higher molecular weight could be introduced for analysis.

At present, measurement of compounds is limited to materials which are unsaturated and have a molecular weight below 378. Modification of the high resolution mass spectrometer by addition of a Field Desorption Source and a second beam would facilitate the introduction and quantitative analysis of high molecular weight material; identification of the compounds observed would be a difficult task requiring a long development program. Nevertheless, the future use of mass spectrometry in studies on liquefaction kinetics depends strongly on characterizing the heavier fractions of the liquid product, which for CO-Steam are the materials that change most markedly during the course of the reaction.

Comparison of both CO-Steam products with Synthoil product in Table 7.2 indicates a distinctly greater extent of hydrogenation for the latter, as would be expected.

The larger amounts of phenols and acenaphthenes/biphenyls indicate a greater breakdown of the macromolecules dissolved from the coal (13); the much smaller quantity of phenanthrenes/anthracenes indicates disappearance of these materials by hydrogenation; and the low concentration of pyrenes along with the high concentration of dihydropyrenes indicates a direct conversion of one to the other by hydrogenation.

The Synthoil product is also characterized by greater alkylation, where a typical aromatic molecule in a Synthoil product contains three alkyl carbon atoms and a typical aromatic molecule in a CO-Steam product contains 0.7 alkyl carbon atoms.

Table 7.2: Comparison of Selected Compound Types in the Volatile Fraction of Different Coal Liquids as Determined by Low Voltage Mass Spectrometry

	Anthracene Oil	CO-Steam Batch Product from GFERC	Lined-Out CO-Steam Continuous Product from PERC	Synthoil from PERC
Volatile fraction, % of total sample	98.6*	78.0*	78.0*	74.0**
Compound types, group total as % of volatiles:				
Phenols	0.0	1.3	8.1	12.2
Indanes/Tetralins	0.0	0.4	3.4	2.3
Naphthalenes	7.7	16.2	3.9	5.1
Acenaphthenes/biphenyls	10.4	6.6	1.5	12.2
Acenaphthalenes/fluorenes	7.1	6.4	5.6	5.0
Dibenzofurans	7.0	7.8	7.5	6.1
Phenanthrenes/anthracenes	19.9	20.2	7.7	2.1
Pyrenes/fluoranthenes	23.4	22.7	9.5	2.5
Dihydropyrenes	0.7	0.7	0.6	9.3
Benzonaphthofurans	2.9	3.3	5.1	3.9
Chrysenes	4.9	3.5	4.0	1.7
Benzopyrenes	3.0	1.9	4.0	1.8

*Determined in mass spectrometer heated inlet at 350°C.
**Estimated from microdistillation data at 250°C and 1 torr.

Source: NTIS GFERC/IC-77/1

Mechanism and the Future of the CO-Steam Process

Data obtained at GFERC indicate that CO-Steam liquefaction in general involves two categories of reactions: (1) reactions occurring over the entire range of temperature investigated (350° to 480°C) which involve a simultaneously rapid consumption of CO and production of a major portion of the total yield of product oil (either soluble oil or distillable oil), and (2) reactions which occur rapidly at or above about 460°C and involve depolymerization of the heavy ends which have been solubilized but not liquefied.

The first class of reactions could be viewed as a rapid solubilization aided by the reducing action of CO and possibly H_2 where the effect of the reducing gas is to remove oxygen from the coal and to enhance the yield of liquid products. Part of the coal material that is rendered soluble in the first category of reactions consists of substances having high molecular weight, which cause the total product to be a solid or a heavy pitch. These heavy materials are fragmented into smaller moieties to an extent depending primarily on temperature and secondarily on residence time in the reactor. To minimize the formation of char by the reverse reaction, the presence of hydrogen donor compounds is essential.

Batch liquefaction with CO and H_2 in 90% anthracene oil-10% Tetralin produces volatile products which are similar to the starting solvent (Table 7.2), and therefore, representative of a pyrolysis-type product. Thus, both the volatile product and the lower molecular weight nonvolatile product appear to be primarily the result of thermal cracking under the conditions investigated, which include temperatures from 400° to 475°C.

Carbon monoxide is an essential reactant for optimum conversion; hydrogen appears less reactive in either category of reaction. It is expected that CO reacts both directly with coal oxygen to form CO_2, and via the formation of "nascent hydrogen." Transfer of hydrogen to the coal fragments depends very importantly on the presence of hydrogen donor solvent during rapid depolymerization.

The future of an economic CO-Steam process should, it is believed, involve higher reaction temperature (above 460°C) and short residence time. This would facilitate rapid conversion to a low-viscosity product, while at the same time avoiding unacceptably high production of methane from recycle solvent. An important part of future work on CO-Steam kinetics is to determine if the CO-Steam product or a fraction thereof possesses hydrogen donor properties which make it a suitable recycle solvent for high temperature operation. In order to reduce the pressure of the process, it may be necessary to consider a product fraction for recycle which contains hydrogen donor materials having a vapor pressure lower than that of Tetralin.

Continuous Processing Unit

The 3- to 5-lb coal/hr Continuous Processing Unit (CPU) shown in Figure 7.8 will occupy a leading position in future studies on the CO-Steam process at GFERC. This unit was designed to facilitate studies on flow and mixing in the reactor for purposes of improving reactor design. The unit will also be used to validate results obtained in the time-sampled batch autoclave, with emphasis on the previously discussed reaction conditions which offer hope for rapid conversion to a relatively low molecular weight product at temperatures above 460°C.

The CPU system (Figure 7.8) consists of:

(1) an input section including the slurry pump and the feed gas compressor;
(2) alternate preheaters (stirred or tubular);
(3) alternate reactors (stirred or tubular);
(4) gas-liquid separation and recycle, including the slurry recycle pump and the recycle gas compressor; and
(5) product accumulators and separators for output of heavy liquid product, light oil, and tail gas.

The ability to substitute either stirred or tubular units for the preheater or the reactor and provisions for gas and slurry recycle provide great flexibility for changing the flow pattern (residence time distribution) and the level of turbulence. This flexibility can be extended by installing baffles in the reactor or by operating reactors in series.

The starting hypothesis in studies on reactor design is that rates of reaction will be enhanced by increasing turbulence and by moving toward the residence time distribution of a tubular reactor rather than a stirred reactor. These general trends have been indicated by experiments conducted at PERC. The systematic study of reactor design parameters for the three-phase CO-Steam process is a challenging task which will require in-depth engineering analysis aided by computer simulation.

Figure 7.8: Schematic Diagram of the GFERC 3-to-5 lb Coal/hr Continuous Process Unit

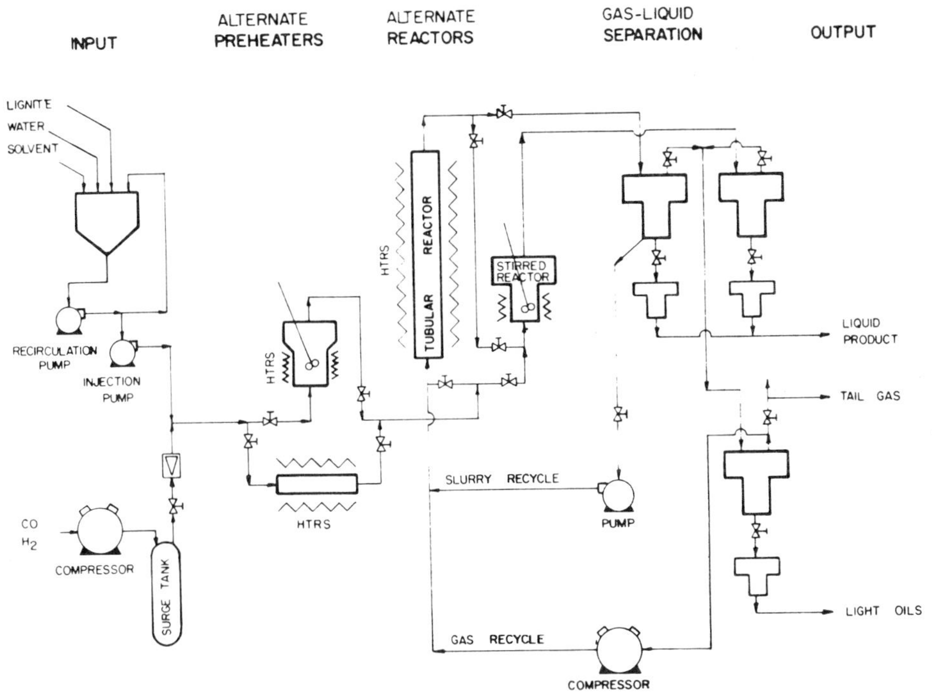

Source: NTIS GFERC/IC-77/1

The CPU is presently undergoing shakedown in preparation for operation on coal in the near future. The first tests on coal will be conducted using a stirred reactor without recycle to establish the operability of the system. The subsequent series of experiments will use a tubular reactor for once through tests to establish optimum reaction rates and product quality along lines investigated previously in the batch reactor.

Process variables to be investigated will include the feed ratios of gas, solvent, and lignite, the extent of slurry preheat, reactor temperature and pressure, and residence time. Thereafter, extended studies will commence on residence time distribution and turbulent mixing.

After the first phase of variable studies are completed on the CPU, work will commence on a planned contract for a feasibility study on the CO-Steam process to address cost competitiveness, the size of a process development unit (PDU) required for scale-up studies, and the possibility of using facilities available now or later from other DOE coal liquefaction projects. In-house work at laboratory scale will be continued concurrently with scale-up.

PROCESS DEVELOPMENT OF LIGNITE LIQUEFACTION

Summary

Progress on the DOE sponsored program for production of solvent-refined lignite (SRL) in a continuous process development unit (PDU) and in laboratory-batch autoclaves is reviewed. Laboratory studies discussed include investigation of process variables such as solvent type and recycle, prior storage, drying, carbonization, composition of hydrogenation atmosphere, lignite source, and particle size of lignite. The unit has been operated from 1,500 to 2,500 psi, with lignite feed rates of 14 to 61 lb/hr and with solvent-to-lignite ratios of 1.33:3.00. Effects of pressure, temperature, and feed gas composition are presented. Economic studies indicated 1985 costs for SRL of \$3.72 per million Btu, assuming 5% escalation and a 15% DCF rate.

Laboratory Studies on Liquefaction

The laboratory studies were conducted in a one-gallon stainless steel stirred autoclave usable to 5,000 psi and 900°F where about 300 grams of moist lignite and 400 grams of solvent were reacted in a reducing atmosphere. Normal operation consisted of heating the autoclave to 750°F at the rate of 5°F/min, and maintaining them at that temperature for 30 minutes. The reaction mixture was then cooled to 400°F, and product gases discharged through cold traps, collected and analyzed.

The autoclave then was repressured with nitrogen and the slurry at 400°F forced through a dip tube onto a heated Buchner funnel where the solvent and lignite-derived liquids were separated from the ash and unreacted lignite. Equipment and operation details, analytical and calculation procedures, and the results of earlier experiments have been published (16).

The following sections present additional batch autoclave results from continued evaluation of process variables for the solution-hydrogenation of lignite.

Solvent Recycle: Fifteen coal-derived and ten petroleum-derived solvents were evaluated at standard test conditions as potential start-up solvents. Details have been published (17). Based on liquefaction performance and availability, a petroleum-derived solvent, Gulf FS-120, a carbon black feedstock, was selected.

To indicate the effect of in-process replacement of the petroleum based solvent by lignite derived solvent, a series of autoclave tests was made in which recovered solvent from previous runs was used in subsequent tests. This sequence of batch experiments was done to simulate a continuous process, and the change in solvent characteristics with successive passes should approach those obtainable in the continuous PDU and would indicate any loss in effectiveness as lignite derived solvent became predominant.

Recycle solvent was recovered by distillation from the solvent-slurry mixture and used for the next pass. Because of handling losses, samples, and the inability to recover all the solvent in the filter cake, three runs at a given recycle level were required to supply sufficient recycle solvent for two runs at the next level. The recycle tests were taken through five passes including a total of forty experiments.

Average yield data are summarized in the following tabulation:

	Pass Number				
	1	2	3	4	5
Number of tests	19	11	6	3	1
Avg. net yields wt % maf lignite					
Net gas	34.0±1.7*	34.1±3.5*	37.1**	33.2**	34.2**
Net liquid	64.5±1.4	64.1±2.1	63.1	64.2	64.4
Unconverted	11.4±0.9	10.9±0.7	10.3	11.2	10.7
H_2O	-10.0±1.0	-9.1±2.1	-10.5	-8.6	-9.6
Net SRL, wt % maf lignite	58.0±1.6	58.2±1.5	56.4	57.9	57.7
Solvent recoverable % of charged	102.0±1.1	102.3±2.4	100.5	99.9	101.9

*The 95% confidence range of the average.
**Confidence range not calculated because of small number of tests.

Effect of Storage on Lignite Liquefaction: Storage is known to alter some properties of coals such as caking, heating value, and moisture content. Consequently, a series of tests was designed to evaluate the effect of various storage conditions on liquefaction of lignite. In most processing applications, particularly development work, some delay between mining and utilization is unavoidable.

A bulk sample of lignite was subdivided, and the representative fractions stored under the following conditions:

(1) Lump lignite under distilled water at ambient conditions. After storage and just prior to liquefaction, the sample was pulverized in a single pass through the mill.

(2) Lignite, pulverized by one pass through the mill, in air at 50% relative humidity, ambient conditions.

(3) Lignite, pulverized by one pass through the mill, in nitrogen at 50% relative humidity, ambient conditions.

After predetermined times in storage, samples from each of the three storage conditions were liquefied using the standard procedures. A duplicate test was made on one of three samples in each set. Storage was continued for 70 weeks.

Pertinent data after 70 weeks of storage are presented below along with baseline data obtained at the start of the series.

	Storage			
	None	Nitrogen	Air	Water
Net yields wt % maf coal				
Gas	35.4	31.3	35.1	33.8
Liquid	66.0	64.5	57.6	66.7
Light oil	20.7	18.1	20.5	22.3
Net SRL	45.3	46.4	37.1	44.4
Unconverted	8.9	9.2	12.4	7.4
Water	-10.3	-5.0	-5.1	-7.9
Conversion, wt % maf coal	91.1	90.8	87.6	92.6

Air storage appears to have a significant adverse effect on lignite liquefaction, i.e., lower net liquid and net SRL yields and lower conversion. Lignite stored in an air-free environment, either in nitrogen or under water, for 70 weeks has about the same reactivity as fresh lignite, as indicated by net liquid and net SRL yields and conversions. Results from liquefaction of samples having intermediate lengths of storage were close to those obtained in baseline tests. The tests series was terminated after 70 weeks.

Influence of Atmosphere Composition: Three different atmospheres, carbon monoxide, hydrogen, and synthesis gas (syngas), were used in otherwise standard batch-liquefaction tests. Syngas was 50-50 mol % hydrogen and carbon monoxide. Initial pressure was 1,000 psig in each case. Both as-received and acid-extracted lignites were tested. Acid extraction (with HCl) lowers the mineral content of the coal and reduces any catalytic effect of the minerals. Yields for these tests are summarized in the following:

	Atmosphere		
	CO	Syngas	H_2
As-Received Lignite			
Net yields, wt % of maf coal			
Gas	66.8	35.2	13.6
Light oil	9.3	13.1	14.5
SRL	33.3	55.3	43.2
Unconverted	17.2	10.0	17.8
Water	-26.6	-13.6	10.9
Acid-Extracted Lignite			
Net yields, wt % of maf coal			
Gas	24.9	27.1	13.6
Light oil	11.3	12.4	14.4
SRL	48.7	47.2	43.4
Unconverted	22.6	12.1	15.8
Water	-7.5	1.2	12.8

Conversion of lignite was generally higher in the syngas atmosphere, 88 to 90%, as compared with 77 to 84% for the other tests.

Yields of liquid products were appreciably higher using the syngas atmosphere and the as-received lignite with its natural moisture content, 68% as compared with 43 to 60% in all the other tests. With acid-extracted lignite the liquid yields were not appreciably different regardless of the atmosphere.

For both extracted and unextracted lignite, yields of gaseous components were higher with more carbon monoxide in the charge gas because of production of carbon dioxide by the water gas shift reaction. The highest gas yield was obtained with carbon monoxide and as-received lignite because of the catalytic effect of the mineral matter.

The syngas atmosphere with as-received lignite gave the best liquefaction, i.e., highest conversion and highest yield of liquid products.

Effect of Drying on Liquefaction: Several advantages have been suggested for drying lignite before liquefaction. Reducing the moisture content would result in lower transportation costs if the liquefaction plant were not at the mine mouth. A lower moisture content of the charged lignite would decrease the generation of carbon dioxide resulting in a reduced load on the gas purification system. Further, the same level of hydrogen-carbon monoxide partial pressure could be maintained at a lower total operating pressure.

The effect of prior dehydration on liquefaction was measured in a test series in which the solvent and lignite were heated at 215°C in the autoclave to remove the moisture. The lignite-solvent mixture was cooled, and a standard liquefaction run made. Tests were made using anthracene oil and FS-120 as solvents.

Reducing the moisture content adversely influenced the extent of liquefaction, especially in the synthesis gas atmosphere. Percentage of unconverted lignite for the dehydrated slurry ranged from 23.9 to 41.6 wt % on a maf basis compared with 8.0 to 9.8 wt % for the baseline tests. Conversion was higher in a hydrogen atmosphere than in the synthesis gas atmosphere. Yields of liquids were low, ranging from 36 to 54.2 wt % of the maf lignite charged, compared to base test liquid yields from 61.0 to 63.4 wt %. Filtration rates for the product slurry were reduced significantly starting with dehydrated lignite. However, yields of carbon dioxide were drastically reduced.

When liquefaction is the primary concern, the gain resulting from reduced carbon dioxide generation will not offset the loss in liquefaction activity resulting from the moisture loss.

In other tests, lignite was dried externally to the autoclave, the dried lignite slurried with the solvent, and water added to the same total water content as when full moisture lignite was charged.

Under standard liquefaction procedures, little difference existed between gas make and conversion. However, yield of light oil was much lower when using dried lignite, and the liquid-solid product slurry was more viscous and difficult to filter, indicating reduced conversion to the lower molecular weight liquids.

Effect of Prior Carbonization: Carbonization or mild heat-treating of lignite prior to liquefaction has been suggested as a means for upgrading lignite, reducing carbon dioxide generated during liquefaction, concentrating methane production, and improving liquefaction characteristics. A number of tests were run in which lignite was mildly carbonized in the presence of the liquefaction solvent or carbonized separately before liquefaction.

Overall, the results from the mild carbonization or heat-treatment of the raw lignite in the presence of the liquefaction solvent show slight deleterious effects on the overall yields and conversion, as indicated in the following table.

	. . . . . Pretreatment Temperature °C			
	None	**260**	**315**	**370**
Yield, wt % maf lignite				
Net gas	38	44	40	33
Net liquid	68	64	61	57
Light oil	15	25	10	12
SRL	53	39	51	45
Net water	-13	-15	-10	-8
Unconverted	7	7	9	18

The effect of carbonization without solvent prior to liquefaction, whether in the autoclave with only gaseous products removed before liquefaction or in a separate carbonizer with only the char charged to liquefaction, is to reduce the liquid yields appreciably:

Carbonization temperature, °C	None	370	375	395
Charge to liquefaction	Lignite	Char and liquids	Char only	Char only
Yields, wt % maf lignite				
Net gas	38	52	28	24
Net liquid	68	35	37	39
Light oil	15	13	4	10
SRL	53	22	33	29
Net water	-13	-15	23	23
Unconverted	7	28	12	14

Although somewhat lower yields of carbon dioxide were achieved and small quantities of methane were produced by carbonization, the undesirable effects of prior carbonization on the liquefaction step make the combined process unattractive.

Effect of Particle Size on Liquefaction: Standard testing procedure involved pulverizing the lignite to -100 U.S. mesh before liquefaction. Previously it had been postulated that such a small particle size would be necessary to insure rapid reaction. One run was made with a larger top size, 0.25 inch, under the otherwise standard liquefaction procedure. The results of this run are compared with the average of three standard runs in the summary following:

Lignite size	-100 mesh	-0.25 inch
Yields, wt % of maf coal		
Net gas	43	45
Net liquid	62	61
Light oil	14	27
SRL	48	44
Net water	-13	-11
Unconverted	8	5

These data indicate liquefaction to be essentially the same under the test conditions employed, either with an initial particle size of -100 mesh or -¼ inch. In pilot plant operation, it should not be necessary to pulverize the lignite completely to -100 mesh to have satisfactory liquefaction since between 100 mesh and ¼ inch the size seems not to be critical.

Effect on Liquefaction of Lignite Source: Some lignites from outside the Fort Union Formation of North Dakota were treated using standard liquefaction procedures in a syngas atmosphere. Lignite from the Denver Basin, Thailand, Chile, and North Dakota are compared in the following.

	Lignite			
	North Dakota*	**Denver Basin**	**Thailand**	**Chile**
Yields, wt % maf lignite				
Net gas	34.0	34.2	31.9	12.6
Net liquid	64.6	65.4	67.0	71.2

(continued)

	Lignite			
	North Dakota*	Denver Basin	Thailand	Chile
Unconverted	11.4	6.5	2.1	6.6
Water	-10.0	-6.1	-1.0	9.6

*Average of 19 tests.

All of the lignites responded adequately to the liquefaction process with no significant difference noted between the North Dakota, Denver Basin and Thailand lignites although the Thailand lignite may be marginally subbituminous. The Chilean lignite responded somewhat differently, not reacting to produce as much gaseous products. This difference appears to be related to the reduced utilization of inherent moisture for a source of hydrogen, probably because the minerals in the Chilean lignite lack the ability to catalyze the water gas shift reaction to the extent exhibited by the minerals in the other lignites.

Process Development Unit

The Project Lignite Process Develpment Unit (Figure 7.9) has a nominal design capacity of 50 pounds of lignite feed per hour and produces approximately 15 pounds per hour of solvent refined lignite (SRL) having a melting point of 300° to 400°F. Additional quantities of lighter liquids and gases are also produced.

Pulverized lignite of full moisture content is slurried with solvent. The slurry is pressurized, preheated and reacted at a selected temperature and pressure in a reducing gas atmosphere. The products are then separated as gases, liquids, and SRL from the unreacted coal and mineral matter. The process is described by function areas in the following.

Coal Crushing and Grinding: The coal is received by either rail or truck. Following initial crushing, a magnet removes tramp iron, and the lignite is split into representative samples using a rotating sampler and placed in storage bins of 65 ft^3 capacity. When required, a bin is brought into the Coal Preparation Building, lifted by an electric crane, inverted, and the lignite delivered through a chute to the crusher. Flow from the bin is controlled using an iris valve. A second iris valve in the chute is used to prevent dusting of coal particles when the bin is removed.

Because of the disparity in capacities of the crusher and pulverizer (3 tons/hr versus 300 lb/hr), a 70 ft^3 capacity storage bunker is directly below the crusher. An emergency dump chute is provided to empty the bunker. Flow from the bunker via a chute to the pulverizer is controlled by an iris valve. From the pulverizer, a cloth chute directs the discharge into a 55-gallon drum for removal to the slurry preparation area.

Slurry Preparation and Pumping: This area includes the pulverized coal bin and the pulverized coal feeder, the slurry mix tank, the slurry circulation pump, and the high pressure reciprocating pumps.

The pulverized lignite is charged to the feed bin of an Acrison volumetric feeder from which it is fed to Marion mixer V-6 at a constant volumetric flow rate. Weigh cells give measurement and control of the weight of coal fed. The flow rate of solvent to the slurry mixer is controlled by a flow recorder-controller

Figure 7.9: Simplified Flow Diagram, 50 lb/hr Process Development Unit, Project Lignite, University of North Dakota

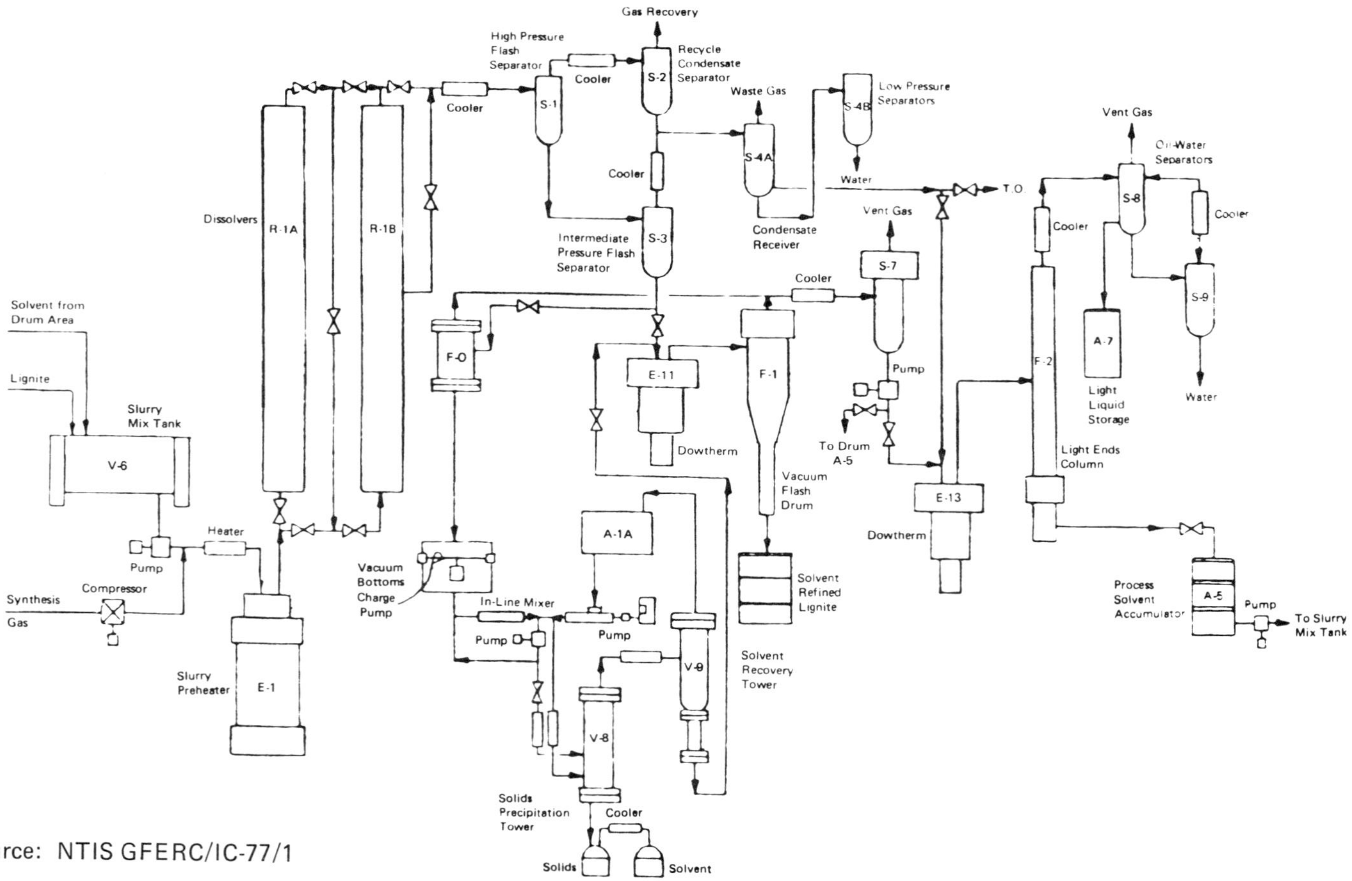

Source: NTIS GFERC/IC-77/1

according to signals from an orifice in the solvent line. The slurry mixer has a jacket through which water or steam can be circulated to control the temperature of the slurry. The flow out of the slurry mixer is controlled by the level of the slurry in the tank.

For circulation of the slurry around the slurry mixer loop, a lobe pump is used which also feeds the slurry through an offtake to the suction of the high-pressure reciprocating pump at about 50 psig. Sample connections in the circulating line are provided so that samples for density and viscosity determinations can be obtained whenever desired. The piping between the slurry loop and the inlet of the reciprocating pump has been made as short as possible to minimize solids settling in the lines.

Slurry Preheating, Dissolving and Gas Separation: This area includes the fluidized solids slurry preheater, the dissolvers, and the high pressure vessels for separation of slurry, liquid, and vapor products. All equipment and vessels in this area have been designed to operate at pressures as high as 2,500 psig and at temperatures to 950°F. The preheater and dissolvers are constructed from Incoloy 800, and the high pressure separators of cast Type 316 stainless steel.

The coal-solvent slurry from the reciprocating pump is combined with a gas stream which is normally composed of recycle gas with some make-up carbon monoxide and hydrogen. This mixture of slurry and gas is heated to about 300°F in a steam-heated exchanger and then enters the coil in fluid solids preheater E-1. In this heater, fine sand is fluidized by air flowing upward in the bed. Thermal requirements are supplied by electric immersion heaters in the fluidized bed and beaded wire heaters on the external walls. The sand is about 50°F above the desired slurry temperature, or generally at 800°F. The mixture of coal-solvent slurry and gas flows through a coil of $\frac{9}{16}$-inch heavy wall Incoloy 800 tubing in the sand bath.

Two dissolvers R-1A and R-1B made from 18-ft lengths of $4\frac{7}{8}$-inches o.d. by $3\frac{7}{8}$-inches i.d. Incoloy 800 tubing are the main reactors. These dissolvers approach adiabatic operation by using controllers reacting to maintain zero temperature gradient between thermocouples at the reactor wall and thermocouples in the insulation. The inlet of each dissolver is at the bottom and the outlet at the top, with another outlet provided at the center. Thus, any multiple of 10 ft high up to 40 ft can be used. Residence times for the slurry can be varied fourfold without changing the feed rate.

An air cooler is used to reduce the temperature of the effluent from the dissolver. The gas-slurry mixture then goes to high pressure separator S-1 at about 550°F. This temperature is adjusted to keep the slurry of coal hot enough to flow but cool enough to reduce polymerization.

The vapor from the high pressure separator is cooled to 100°F or lower in a water cooler and then enters recycle condensate separator S-2. The vapor products from this separator consist primarily of unreacted carbon monoxide and hydrogen, carbon dioxide, hydrogen sulfide, and light hydrocarbons. The gas passes through a pressure control valve and into the gas recovery and recompression area, where it is purified and recycled. The pressure control valve controls the pressure on the entire preheater-dissolver-high pressure separator system. The liquid products from the recycle condensate separator consist mainly of water, process solvent, and lighter products. This stream goes to low pressure flash separator S-4A.

The hot slurry products from the high pressure separator flow through a captive volume pressure letdown system and then enter intermediate pressure flash separator S-3 which is maintained at the same temperature as the high pressure separator. The vapor products from this separator consist of light hydrocarbons and some solvent. This vapor passes through a pressure control valve and is then cooled and condensed to 100°F or below by water coolers. This cold stream combines with the liquid products from the recycle condensate separator to enter low pressure flash separator S-4A.

In the dual low pressure flash separators S-4A and S-4B, the aqueous phase is separated from the organic phase. The organic phase goes to the light ends column feed heater.

The slurry from intermediate pressure flash separator S-3 flows through a second captive volume pressure letdown system. Rate is controlled by a displacement type level controller within the vessel, and the stream then goes to the partial flash separator F-0. The overhead from F-0 is combined with the overhead from vacuum flash drum F-1 which go through a condenser to vacuum condensate accumulator S-7. The F-0 bottoms goes to solid-liquid separations.

Solid-Liquid Separation: Mineral separation consists of a system of vessels and pumps designed for the high temperature extraction of solvent refined lignite plus solvent from the mineral matter and unconverted lignite using toluene as a diluent. The flow diagram for this area is shown in Figure 7.10.

The bottoms stream from partial vacuum flash drum F-0 is pumped through an in-line mixer into a recirculating slurry loop along with the toluene deashing solvent.

The toluene-slurry lobe pump discharges the slurry at pressures to 450 psig. The slurry is fed through an in-line, cyclindrical, 5 kVA heater and the process temperature is raised to 350° or 400°F. The mixture is then fed to the settling tower V-8. This tower is 18 inches in diameter by 12 feet high and is fabricated of carbon steel. Sufficient cross-sectional area is provided so that the terminal velocity of the settling particles is greater than the upward velocity of the toluene and SRL. Particles of one micron or larger size will readily settle through the toluene solution moving upward at a velocity of 0.005 foot per second or less.

At the base of the settling tower, a second stream of toluene enters at the temperature and pressure of the tower. This flows upward, promoting dissolution of any SRL which may be adhering to the ash and unconverted lignite. Mineral matter, unconverted lignite, and some toluene are withdrawn through a sort of lock hopper operated intermittently by a timer. The toluene is flashed, condensed and collected. The ash and unconverted lignite are collected in a drum.

The liquid overhead from the settling tower, consisting of dissolved SRL, liquefaction solvent, and toluene, is depressurized, cooled in the steam cooler and fed to the toluene recovery tower, V-9. This tower is 6 inches in diameter by 10 feet and operates at 300°F or above and 5 psig. The toluene flashes, leaving the SRL-solvent mixture behind. The bottom stream is fed to preheater E-11 for vacuum flash separation of SRL and liquefaction solvent.

Figure 7.10: Solid-Liquid Separation, 50 lb/hr Process Development Unit

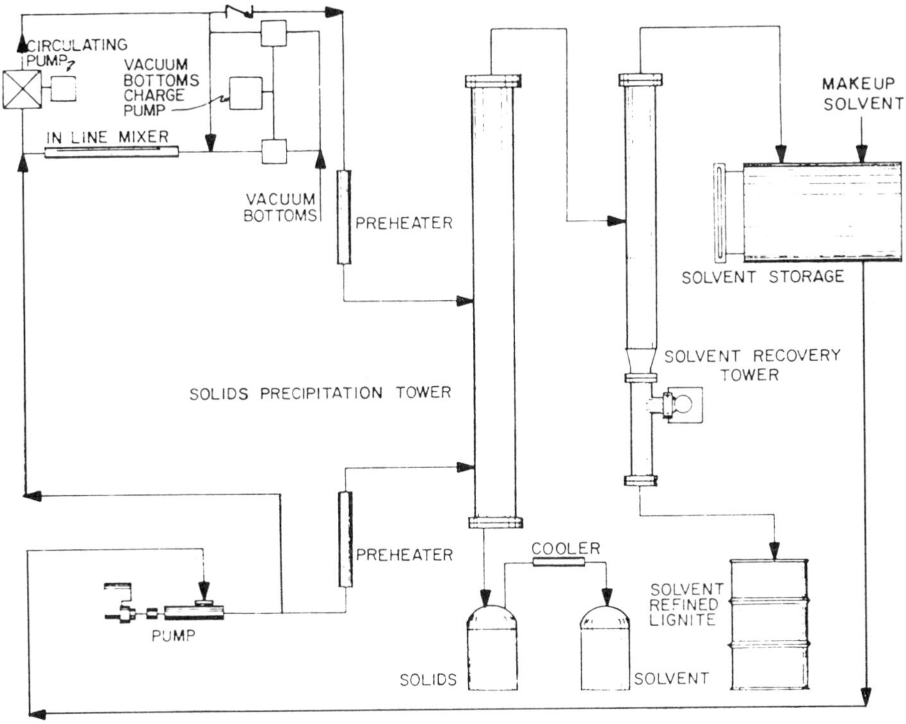

Source: NTIS GFERC/IC-77/1

The toluene is taken overhead, condensed and stored in A-1A accumulator. This recycle toluene is fed to the suction of a triple diaphragm pump. The discharge stream of toluene, at a pressure up to 450 psig, is split into two streams. One provides feed toluene for dilution of the vacuum bottoms and the second feeds toluent through a line preheater to the wash toleuene inlet on settling tower V-8.

Liquid Separation and Solvent Recycle: The two major distillation sections located in this area are the vacuum flash section for separating process solvent from the liquefied coal product and the distillation column for separating light oil from recycle liquefaction solvent. The Dowtherm preheaters, storage tanks for all products, and solvent recycle facilities are also included in this area.

The slurry from the intermediate pressure flash separator S-3 at about 550°F goes into flash drum F-0. The bottoms stream from this flash drum goes to the solid-liquid separations system. The return stream from the solid-liquid separations system, i.e., the bottoms from the toluene flash vessel V-9, is fed through preheater E-11 and into the vacuum flash drum F-1. The bottoms from F-1 is the SRL product. The overhead from F-1 is combined with the F-0 overhead, condensed, and sent to drum A-5A as part of the recycle solvent. The light oil from S-4A is charged to light ends column F-2. This column is operated at a pressure of 10 to 15 psig. The overhead product is a light oil, essentially a stabilized naphtha, consisting of hydrocarbons ranging from about C_5 to perhaps as high as 350° to 450°F in boiling point. The bottom product also is recycled through drum A-5A to the slurry mix tank.

Gas Purification and Recycling: The vapor products from the recycle condensate separator S-2 at about 100°F flow through a pressure control valve in which the pressure is reduced to about 100 psig to an absorber tower in which carbon dioxide and hydrogen sulfide are removed using a 10% by weight solution of sodium hydroxide. The upper two-thirds of the tower is packed with 1-inch polypropylene Pall rings on which sodium hydroxide solution is sprayed. The gas percolates upward through this sodium hydroxide solution. The sodium hydroxide from the bottom of the absorber tower flows back to the storage tank through a level control valve, operating on the level of sodium hydroxide solution in the tower. The sodium hydroxide is circulated until it is spent, and then a fresh solution is charged from the other storage tank.

The gas stream from the acid gas absorber tower goes to the ammonia absorber tower containing boric acid solution. The gas from the ammonia absorber tower consists of carbon monoxide and hydrogen, plus light hydrocarbons and inerts. Part of the gas flows to a compressor surge drum through a positive displacement meter and a bleed stream flows to the thermal oxidizer through a flow control valve to limit the build-up of hydrocarbon gases and inert gases.

Two identical diaphragm compressors are provided, either of which has sufficient capacity for the requirements of the unit.

The feed hydrogen and carbon monoxide are stored in high-pressure tube trailers. The ratio of hydrogen to carbon monoxide in the feed is controlled by a ratio recorder controller. The hydrogen-carbon monoxide gas mixture is combined with the recycle gas just ahead of the suction of the compressor, with the flow rate controlled to maintain the desired suction pressure. After compression the gas flows through a flow controller to the slurry just ahead of the fluidized solids preheater E-1.

Materials of Construction: Considerable effort was spent during the design of the PDU on selection of materials of construction so as to avoid major corrosion problems while still providing data suitable for scale-up. Type 316 stainless steel was selected for areas where design conditions were 3,000 psi and 950°F so as to avoid hydrogen embrittlement which was believed to be the major corrosion mechanism. Beyond the high pressure section, type 304 stainless steel was generally used. Type 316 stainless steel had been used for years in batch autoclave work with lignite. Further, no major corrosion problems had been noted at other locations in continuous liquefaction operations using bituminous coal.

After 400 hours of preliminary operation, including approximately 125 hours with lignite in the system, the preheater coil constructed from $^9/_{16}$" o.d. x $^5/_{16}$" i.d., type 316 stainless steel tubing plugged. During removal of the plug, the coil fractured near the slurry inlet. Two replacement coils of 316 stainless steel failed during later operation. One failed after 250 hours of operation and the other after only 28.

Samples cut from the coils were examined and branched-transgranular stress-corrosion cracks were found that initiated at the inside of the tube. Metallurgical examination diagnosed the failures as classical examples of chloride induced stress corrosion. Little evidence of other corrosion was evident. In the course of additional operation and investigation, stress corrosion cracking was found in the dissolvers, and in tubing and fittings in high-pressure, high-temperature areas.

Samples of the type 316 stainless steel were anlayzed and found to meet specifications. Chloride concentrations of 536 and 379 ppm were found by water extraction of the inside of a preheater coil section. Sodium ion concentrations were approximately a tenth of the chloride concentration. It was believed that the chloride at the extract pH of 5 to 6 would be sufficient to initiate the stress corrosion cracking.

Possible sources of chloride were investigated. The Gulf FS-120 feed solvent contained about 90 ppm chloride, the recycle solvent near 200 ppm, and the lignite feed 40 ppm. Such concentrations are not high by comparison to those in similar liquefaction processes using coal of high rank where this type of corrosion has not been a problem. The very rapid failure of the third coil was most unexpected. It was suggested that perhaps the chloride would be concentrated at the interface where the moisture was evaporated from the lignite which would be near the inlet to the coil preheater. A chlorinated solvent had been used to clean various components in the low pressure sections of the PDU including the slurry-mix tank.

Although care was taken to remove the solvent, 1,1,1-trichloroethane, after cleaning, the design of the mix tank was such that complete residue removal was difficult. The slurry mix tank was replaced with a trough mixer and cleaning with chlorinated solvent was discontinued.

Because of the major stress corrosion problems with the stainless steel, the entire high-pressure, high-temperature section of the PDU was replaced with Incoloy 800. This replacement resulted in about a six-month delay in operations, and the wall thickness of the Incoloy 800 tubing available limited the maximum operating pressure to 2,500 psig.

The PDU has been operated from January of 1970 through April of 1977 without appreciable corrosion of the Incoloy. Microexamination of the tubing samples after a month's service indicated the presence of shallow surface cracks. However, an unused section of tubing exhibited the same appearance.

One other experience with corrosion necessitated a change in materials. The overhead line from the primary separator to the condenser just ahead of the second high-pressure separator was penetrated by a pinhole apparently resulting from acid attack where water and light oils condensed. The original type 304 tubing was replaced by type 316 tubing and there have been no more apparent problems in this area.

Possible corrosion problems in the PDU continue to receive considerable attention. Metal samples are periodically removed for microscopic examination. No stress corrosion attack has been noted in the stainless steel (type 316) fittings replaced in January 1976. However, corrosion of thermocouple wells and at piping bends has been noted.

Typical PDU Operation: The PDU is normally operated continuously for a five-day week and data gathered for a number of twelve-hour yield periods for material balance purposes. In Figure 7.11 is shown a flow schematic for Run M-17AB, a composite of two yield periods run back to back. The mass flow rates in and out and in recycle streams are shown as well as pressures and temperatures in important vessels. In this run, 41 lb/hr of 31.5% moisture lignite was processed with 536 scfh of gas containing 75% H_2 and 25% CO, 81 lb/hr of recycle liquefaction solvent, and 107 lb/hr of toluene recycle in the liquids-solids separations system.

Feed and product compositions and characteristics are shown in Table 7.3, and yield structure shown as a bar graph in Figure 7.12. In the basis of 100 lb of maf lignite fed, 65.8 lb of SRL and light organic liquids were produced, and 15.6 lb of CO, 1.0 lb of H_2 and 5.4 lb of water were consumed. Conversion was a little more than 90% of the maf coal.

Table 7.3: Feed and Products—Run M-17AB

Lignite
North American Coal Co., Zap, N.D.
Screen size 90% -200 mesh, 100% -60 mesh
Moisture: 31.5%

Feed gas (mixed from tank trailers)

	Vol. %	Wt. %
CO	24.7	81.9
H_2	75.2	17.8

Startup liquefaction solvent
Gulf Oil Co. FS-120 Carbon Black feed stock
Boiling range (1.6 torrs)

	FS-120 Startup	Recycle Solvent
IBP-100°C	4.2	34.9
100°-230°C	81.0	55.5
230°-255°C	14.8	9.6
Specific gravity	1.046	0.980

Material balance for gas components

	Vol. % In	Vol. % Out	lb/hr In	lb/hr Out
H_2	75.2	66.2	2.13	1.87
CO	24.7	14.4	9.78	5.81
CO_2	–	13.2	–	9.91
H_2S	–	0.2	–	0.18
CH_4	–	4.5	–	1.11
C_2H_6	–	1.1	–	0.58
C_3H_8	–	0.3	–	0.33
NH_3	–	0.1	–	0.04

Ultimate analysis of materials

	Lignite Charged	Starting Solvent	Recycle Solvent	Vacuum Bottoms	Deashed SRL
C	45.22	89.03	83.60	80.20	87.42
H	6.43	8.11	9.14	5.20	5.67
N	0.64	0.12	0.20	0.98	1.07
S	0.45	2.23	1.09	0.90	0.98
O*	41.45	0.51	5.97	4.46	4.86
Ash	5.81	0	0	8.26	0

*By difference.

Properties of product SRL

	F-1 Vacuum Bottoms (measured)	SRL (calculated)
Gradient bar melting point, °F	327	–
Pyridine solubles, wt % ash-free	83.5	100
Specific gravity	1.28	1.25
Heat of combustion, Btu/lb	14,330	15,990

Source: NTIS GFERC/IC-77/1

Figure 7.11: Mass Rates and Run Conditions

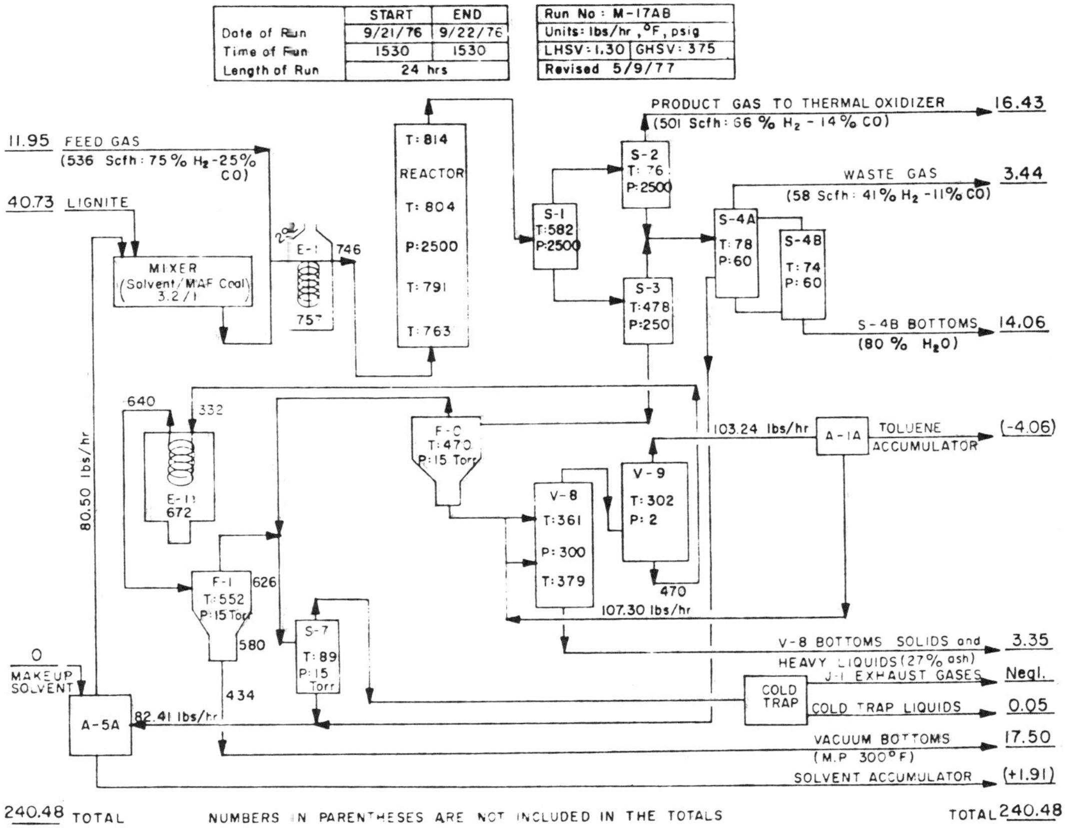

Source: NTIS GFERC/IC-77/1

Figure 7.12: Net Material Balance, Run M-17AB

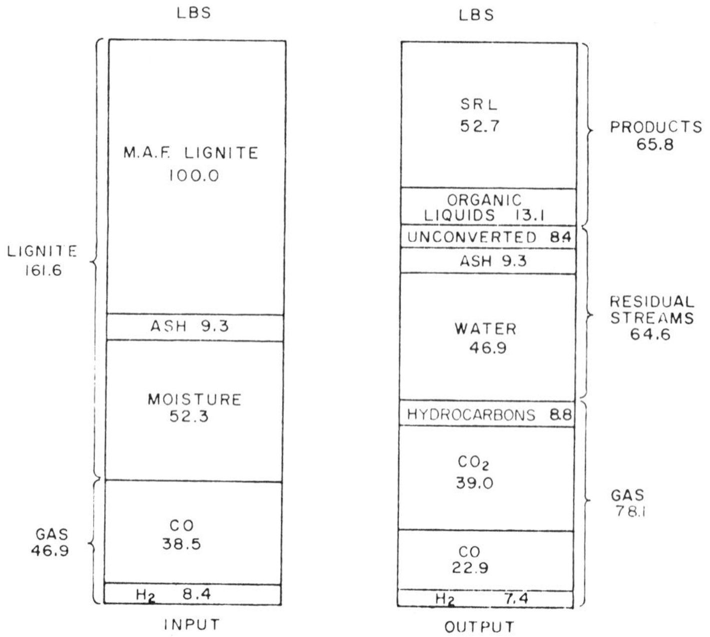

Source: NTIS GFERC/IC-77/1

Cost Estimates for a Commercial SRL Plant

The commercial version of the solvent refined lignited (SRL) process is based on the report Demonstration Plant Clean Boiler Fuels from Coal—Preliminary Design/Capital Cost Estimate prepared by the Ralph M. Parsons Company, Office of Coal Research, R & D Report No. 82—Interim Report No. 1 (1973). The SRL Plant would use 29,328 tons of lignite per stream day and produce 7,200 tons of SRL and 11,000 barrels of light oil. The gross heating value of the SRL and light oil totals 298.8 billion Btu per day. Major modules of the plant are coal liquefaction and filtration, gasification, oxygen production, and purification.

Lignite is crushed, ground and combined with recycle solvent to form a slurry. The slurry is charged to a vessel where it reacts with a reducing gas composed of recycle gas and make-up synthesis gas. The product of this reaction passes to a filtration unit which separates the product into a filtrate which is sent to distillation and a wet filter cake which is conveyed to the gasification unit. The filtrate is distilled to recover solvent and to produce SRL and light oil, the two main products. Wet filter cake from the liquefaction process is used for the production of synthesis gas. Gases produced in coal liquefaction, distillation, and gasification are sent to acid gas removal where carbon dioxide and hydrogen sulfide are removed.

Carbon dioxide is vented and hydrogen sulfide moved to sulfur removal where a purification unit recovers the sulfur and produces a tail gas which is also vented to the atmosphere. The purified gas from gasification is used mainly as a reducing gas by the liquefaction process and the balance is used with other purified gases as plant fuel.

The total capital required is $602 million, consisting mainly of plant cost shown in Table 7.4. Total plant cost is on an insurance and tax basis and consequently consists of all direct and indirect cost of plant construction. Start-up cost is taken as 3% of total plant cost. Return on equity is the charge for use of stockholder funds for financing plant investment. The data in the table are based on a 15% rate of return. Working capital is taken as 9% of the subtotal for depreciation.

Cost for lignite is the largest operating item. The plant uses 9.7 million tons of lignite annually or 29 thousand tons per day at $4.00 per ton. Eighty-seven direct laborers are needed daily to operate the plant and are hired at $6.00 per hour. Twice as many maintenance laborers as direct laborers are needed and are paid a yearly salary of $15,000. The cost of maintenance material is taken as 4% of the total of plant cost and start-up cost. Supervision is 15% of the cost of direct and maintenance labor; payroll overhead is 40% of payroll. Other direct production costs consist of the costs of chemicals, catalysts, power, and water. Taxes and insurance are 2% of total plant cost.

Administration and general overhead are 40% of direct and maintenance labor and supplies; supplies themselves are 20% of maintenance materials and labor. Depreciation is calculated by the sum-of-the-years digit method over the first 16 years of the project and is not shown in Table 7.4. However, the 16-year average of depreciation is $31.25 million per year.

Table 7.4: Economic Summary (Mid 1976 Prices)

	Capital Requirement, Millions of Dollars
Total plant investment	500.00
Depreciation subtotal	500.00
Return on equity during construction	42.50
Start-up costs	15.00
Working capital	45.00
Total capital requirement	602.50
Annual operating costs	
Lignite (1,222 tons/hr @ $4.00/ton)	38.70
Direct operating labor ($6/manhour)	1.50
Maintenance labor and material	23.20
Supervision and payroll overhead	2.30
Other direct production costs	6.80
Taxes and insurance	10.00
Administration and general overhead	11.80
Total annual operating cost (excludes depreciation)	94.30

Source: NTIS GFERC/IC-77/1

Investment is assumed to occur on a 11:33:56 basis over a 3-year period and to be 100% equity financed. The sum-of-the-years digit method is used to depreciate the investment over 16 years to the 20 years of operation. Given these parameters, the cost per unit of output for a discounted cash flow (DCF) rate of 15% is $2.46 per million Btu.

Assuming different years of start-up (1980, 1985, 1990) and different rates of escalation (3, 5, and 7%) to start-up, plant investment increases as follows:

Escalation Rate Percentage	Plant Investment (millions of dollars) ....Year of Start-Up.......... 1980	1985	1990
3	554.5	642.8	767.6
5	593.1	757.0	1,014.4
7	633.6	888.6	1,333.6

Thus, plant investment could be 11 to 27% higher in 1980 depending on whether inflation will be at 3 or 7%. If inflation is at a moderate rate of 5%, plant investment doubles in 14 years. Sensitivity of unit cost to changes in the DCF rate, year of start-up, and escalation rate is reflected by the figures shown in Table 7.5.

Table 7.5: Sensitivity of Unit Cost to DCF Rate, Start-Up Year, and Escalation

	Unit Costs, $/MM Btu ....DCF Rate..........		
	10%	15%	20%
1976 Base (no escalation)	1.96	2.46	3.02
3% Escalation to:			
1980	2.12	2.72	3.35
1985	2.52	3.16	3.19
1990	3.00	3.77	4.58
5% Escalation to:			
1980	2.32	2.91	3.59
1985	2.97	3.72	4.58
1990	3.97	4.98	6.13
7% Escalation to:			
1980	2.48	3.11	3.83
1985	3.48	4.36	5.37
1990	5.22	6.55	8.06

Source: NTIS GFERC/IC-77/1

FLASH HYDROPYROLYSIS PROCESS FOR CONVERSION OF LIGNITE

Introduction

Brookhaven National Laboratories (Upton, N.Y.) is developing a Flash Hydropyrolysis process to produce 35% light hydrocarbon gases (methane and ethane), 25% liquids (15% benzene and 10% oil) and 40% chars from a North Dakota lignite. The process converts lignite to liquid and gaseous hydrocarbons, directly, by rapidly heating lignite with preheated hydrogen to reaction temperatures followed by rapid cooling. The Flash Hydropyrolysis process offers flexibility in either production of aromatic and olefinic chemical feedstocks or unleaded gasoline production from coal.

Process Description

In the Flash Hydropyrolysis process (18) powdered pressurized coal is fed together with hydrogen into a quasi-entrained flow reactor where lignite and hydrogen react in a relatively short residence time to produce benzene and other light hydrocarbons. The system is maintained under 1,500 psig and 1300°F. The process scheme is shown in Figures 7.13 and 7.14.

The gas and solid are separated in multistage cyclones and the char then routed to the BCR gasifier for hydrogen production. The gas stream from the gas-solid cyclone is then passed through an iron oxide tower for H_2S removal before being directed through a hydrogen stream via recycle hydrogen heat exchanger. A portion of the gas stream is recycled while about one-half of the gas stream is passed through a quench tower. BTX produced in the flash hydrogenation reactor is then quickly quenched with a recirculating oily-liquid stream and the product liquid leaves the liquid separator to go to an extractive distillation unit such as a Udex or Sulfolane unit for benzene extraction. Light liquid is sent to a stabilizer to yield light oil which is similar to No. 4 fuel oil.

Overhead hydrocarbon vapor product from the quench tower is passed through MEA and copper wash units, i.e., Cosorb to remove all residual sulfur, CO_2 and CO before being directed to a gas separation unit where methane and ethane are cryogenically separated and recovered from the hydrogen which is recycled back to the flash hydropyrolyzer. CO recovered in the copper wash is fed to the hydrogen furnace as fuel.

The gas separation system is essentially a combination of low-temperature condensation and absorption processes, in which the bulk of the C_2 and heavier is condensed out as the gas is cooled. The remaining C_1 is removed from hydrogen in an absorber-stripper system in which liquid propane is utilized as the absorption medium and a portion of the product C_1 as stripping gas. Propane is selected for this application because of its lowest freezing point among the paraffinic hydrocarbons (-306°F) and having a boiling point of -43.8°F, which is more than 200°F higher than C_1, the light key component to be separated from the propane. Make-up propane may be added to compensate for the small quantity lost in the stripper. C_3 is obtained from the ethylene plant where the recovered ethane is cracked to make ethylene and other products.

The feed gas to the gas separation plant contains 85% hydrogen, 10% methane, and 3.4% ethane. The gas stream is flashed from 1,500 to 650 psia at a temperature of 250°F. The flashing effect brings the stream temperature down to 84°F. The feed then is passed through a drying unit, where the water dew point is reduced to -100°F. The stream then flows through a bank of the first multistream extended surface exchangers and is further cooled to -120°F. After the first separation, the gas is cooled in progressive steps to -160°, -200°, -270°F.

Ethane and heavier components are condensed in the first condenser. Residual C_2 and heavier are removed in the second condenser as liquids. Ethane recovered is sent to the ethylene complex and cracked to produce ethylene, propane, butadiene and aromatics.

Figure 7.13: Flash Hydropyrolysis of Coal to Chemical Feedstocks and Pipeline Gas

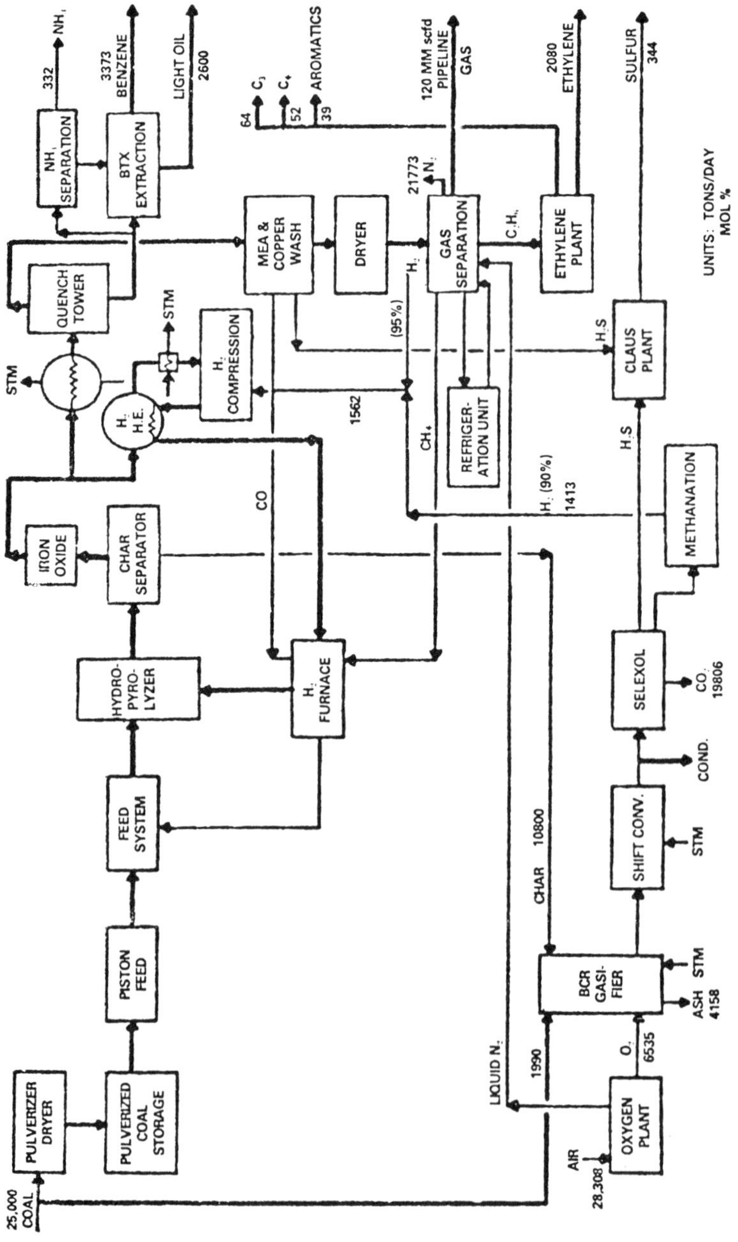

Source: ERDA CONF-760401-5

Figure 7.14: Flash Hydropyrolysis of Coal to Gasoline and Motor Distillates

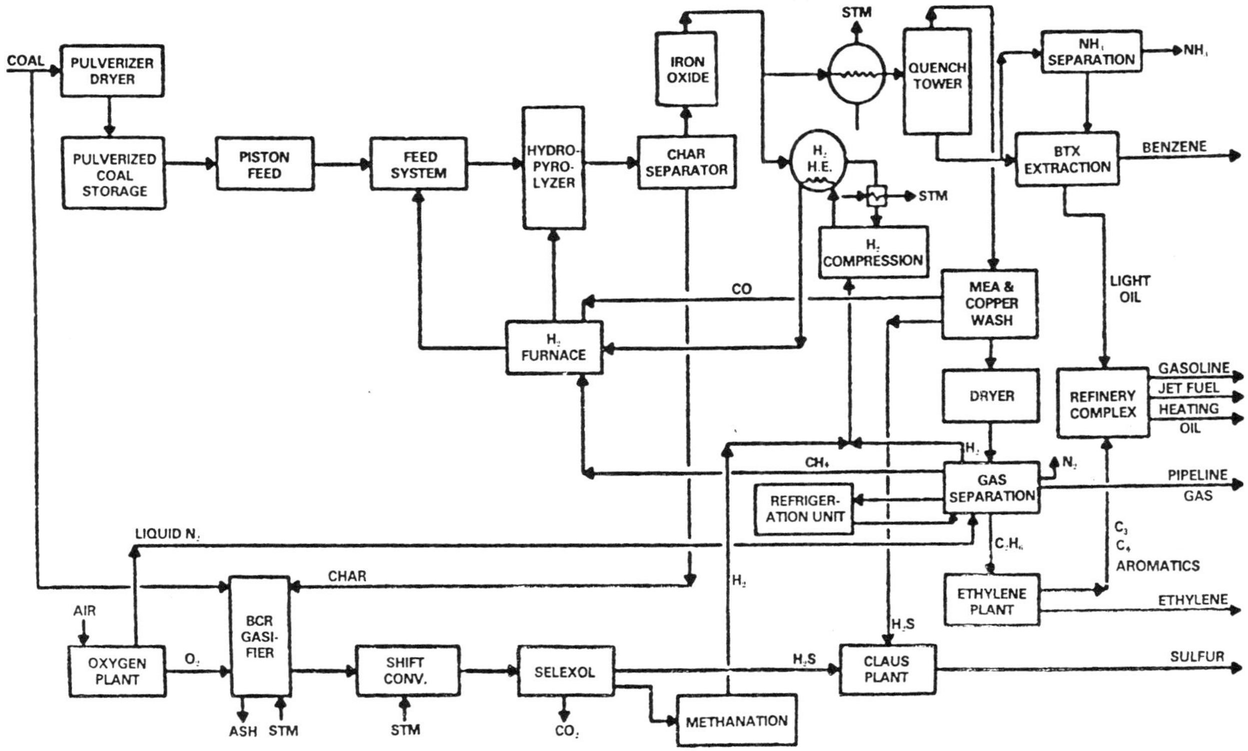

Source: ERDA CONF-760401-5

The major portion of the vapor overhead from the first condenser is C_1 which is sent to the battery limit and recompressed to bring it up to pipeline gas pressure.

The return hydrogen from the gas separation plant is scrubbed in the methane absorber tower in which propane is used as the scrubbing medium. Residual hydrocarbons are also removed by countercurrent washing with liquid propane. The rich propane is then pressurized, heated, stripped of C_1, and recycled back to the methane absorber.

About 10% of the total heat removal requirement for cryogenic separation is supplied by the auxiliary refrigeration unit. The necesssary cryogenic heat load is supplied by an ethylene and propylene cascade system.

The ethylene unit consists of the following major processing areas:

> The Pyrolysis Area—where ethane feedstock is pyrolyzed with steam to produce olefins and by-products. It includes waste heat recovery and supplemental steam generation.
>
> The Quench and Compression Area—in which the hot pyrolysis gas is cooled and compressed to feed the purification and recovery steps.
>
> Purification and Recovery Area—where the cool, compressed product is purified, dried, and fractionated into the various products and by-products.

Temperature reduction can be accomplished by a number of direct-contact and indirect-contact methods. With ethane feedstock, transfer-line heat exchangers reduce by a substantial amount the heat that would be otherwise delivered to the cooling tower.

Absence of heavier feedstocks to the ethylene plant allows for the design of a very simple olefin plant.

The BCR gasifier is fed with char from the hydropyrolysis unit, together with additional make-up coal. Oxygen required for gasification comes from an adjacent air separation plant. The liquid nitrogen produced is utilized to supplement a part of the refrigeration unit load.

The gases from the gasifier are converted and treated in a shift converter, a Selexol unit for acid gas removal, and a methanation unit for carbon monoxide conversion. The H_2S and CO_2 are removed in the Selexol unit, combined with acid gas recovered in the other section, fed to a Claus plant and converted to elemental sulfur. Tail gas processes are included here to insure the EPA emission standards are complied with.

An alternate-scheme to make gasoline and motor distillates is shown in Figure 7.14. Light oil and part of the C_2 and heavier refinery liquids are processed in a refinery complex for production of gasoline, jet fuel, and fuel oil.

In Tables 7.6 and 7.7, material and energy balance and the overall weighted utility requirements are shown for an integrated coal complex utilizing the Flash Hydropyrolysis Process to make chemical feedstocks and pipeline gas.

Table 7.6: Flash Hydropyrolysis Process—Material and Energy Balance

	Tons/Day	MM Btu/Day
Feed		
Lignite to Flash Hydropyrolyzer	25,000	568,400
Lignite to BCR gasifier	1,990	45,244
H_2O for plant (net)	11,244	
O_2 for H_2 generation	6,536	
Total	44,770	613,644
Products		
Benzene	3,375	121,587
Light oil	2,600	103,740
Hydrogen	1,562	
Ethylene	2,080	85,246
C_3	64	2,588
C_4	52	2,145
Aromatics	39	1,436
Pipeline gas	4,828	216,089*
CO	1,292	
CO_2	20,015	
NH_3	332	
Sulfur	344	
Ash	4,158	
Water	3,953	
Cresol, xylenols	76	
Total	44,770	532,831

Note: Overall thermal efficiency is 86.8%; net thermal efficiency is 71%, both based on net production of total products.

*120,024 net product

Source: ERDA CONF-760401-5

Table 7.7: Flash Hydropyrolysis Process—Weighted Utility Summary

Unit	Cooling Water (gpm)	Electricity (kWh)	Steam (lb/hr)	Fuel (MM Btu/hr)
Lignite preparation and feeding system	1,200	15,300		
Char separation		200		
Quench tower	103	4,478	–(772,325)	
BTX extraction	15,927	1,462	244,369	9
Light oil unit	8,171	420	4,333	80
Oxygen	39,210		414,318	
BCR gasifier		9,011	733,000	
Shift conversion			–(264,000)	
Cosorb	11,345	4,245	189,004	
Selexol	20,000	37,285	250,000	
Sulfur plant	122	210	–(73,710)	
H_2 compression			228,780	
Gas separation	9,654	14,914	114,060	
Ethylene unit	44,100		–(255,900)	2,320
Refrigeration	5,870			

(continued)

Table 7.7: (continued)

Unit	Cooling Water (gpm)	Electricity (kWh)	Steam (lb/hr)	Fuel (MM Btu/hr)
Hydrogen furnace				2,592
Hydrogen steam generator			−(2,433,558)	
Total	155,702	87,525	−(1,621,629)	5,001

Note: The cooling water, electricity, steam, and fuel demands and production rates were modified for use in the above listed process units to achieve optimum utilization of the available energy. Figures in parentheses indicate quantities produced and available for utilization in other process operations.

Source: ERDA CONF-760401-5

Economic Considerations

The investment for the Flash Hydropyrolysis Process to make chemical feedstocks is summarized in Table 7.8. The summary of the manufacturing costs and revenues is presented in Table 7.9. These annual operating costs at full production are based on 90% service factor, except for the first year of plant operation which is assumed operating at 50% of full capacity.

Depreciation for the coal complex is set at 5% (20-year life) on a straight line basis. The profitability of the complex was evaluated using both a discounted cash flow (DCF) return on investment and a simplified annual (after-tax) return on investment (ROI, Table 7.9). Investment costs for the plant were determined by a detailed estimation of the major process units on a modularized basis. Investment period and profile are assumed for a completion time of three years, from detailed engineering to start-up. The percentage of investment expended per year is taken as 20% for year 1, 45% for year 2, and 35% for year 3.

The cost of the land is set at $5,000 per acre for uniform comparison purposes. Working capital is based on 60 days inventory cash supply. This cash is to be financed in the 4th year of the project.

Table 7.8: Modularized Installed Cost of the Flash Hydropyrolysis Process*

Unit	$ MM
Lignite preparation, storage, transport pressurizing and feeding systems	67.87
Hydropyrolyzer	33.52
Char and solid separation system	6.71
Iron oxide tower	2.31
Quench tower	2.96
BTX extraction unit	15.60
Dryer	1.60
Cosorb and MEA system	35.00
Gas separation unit	24.70

(continued)

Table 7.8: (continued)

Unit	$ MM
Ethylene plant	83.74
H_2 furnace	24.60
H_2 recycle/superheat steam generator	4.87
H_2 recycle compression	39.00
H_2 feed compression	13.50
O_2 plant	46.00
BCR H_2 generator	22.50
Shift converter	40.00
Methanation	9.00
Selexol	36.50
Claus plant	4.00
Refrigeration unit	5.23
Light oil stabilizer	2.67
Subtotal	521.88
Services	42.41
Site preparation	8.00
Land	2.00
Total	574.29

Note: Mine cost considered as an equity; equity/debt ratio 20/80. Ownership of mine assumed.

*25,000 ton/day lignite throughout

Source: ERDA CONF-760401-5

Table 7.9: Flash Hydropyrolysis Process—Economic Evaluation

Investment $574.29 MM*

Income

Product	Quantity		Price		$/Day
Pipeline gas	120	MM scfd	1.5	$/MM Btu	180,000
Ethylene	1.37	billion lb/yr	10	¢/lb	416,000
Propylene	128,000	lb/day	30	¢/gal	8,847
Butadiene	104,000	lb/day	16	¢/gal	16,416
Aromatics	78,000	lb/day	75	¢/gal	7,942
Benzene	21,815	bbl/day	75	¢/gal	732,984
Light oil (No. 4)	18,934	bbl/day	11	$/bbl	208,274
Ammonia	332	ton/day	150	$/ton	49,800
Sulfur	344	ton/day	25	$/ton	8,600
Electricity	79	MW	15	mills/kWh	28,440
					$1,657,303

Note: At 330 days operation (7,920 hr/yr) 90% c.f. — yearly income = $546.9 MM

Expense

	MM $/yr
Lignite at 65¢/MM Btu	158.50
Catalyst and chemicals	5.00
Power at 15 mills/kWh	21.15
Raw water at 5¢/M gal	0.38
Ash disposal at $1.20/ton	1.74
Insurance at 2% of capital investment	10.69
Maintenance at 4% of capital investment	21.38

(continued)

Table 7.9: (continued)

Expense	MM $/yr
Operating labor at $15,000/yr	5.16
Supervision at 15% of labor cost	0.77
Administration and support labor at 20% of all labor cost	1.19
Payroll burden 20% of all labor	1.19
General administrative expenses at 2% of the facilities investment	10.69
	237.84
Depreciation at 5% (20 yr)	28.71
Mortgage at 10%	69.56
	336.11
Income before taxes	210.80
Net profit (50% of income)	105.40
ROI	18.35
DCF after tax return on investment	10.50

*Mine cost considered as an equity; equity/debt ratio 20/80

Source: ERDA CONF-760401-5

The major income for the coal complex is derived from benzene, ethylene, and pipeline gas. A larger recycle hydrogen stream will significantly lower pipeline gas production. This is due to the reheating requirement of the recycle stream, thus increasing methane consumption which would otherwise be diverted to pipeline gas production. On the other hand, a smaller recycle hydrogen operation will enhance the rate of return on the investment. The simple ROI (after-tax) for the chemical refinery is 18.35% and on an annual return and DCF it is 10.5%, which is economically reasonably attractive.

The estimate for a flash hydropyrolysis plant making motor distillate and aviation fuel only indicates less favorable economic return than for an integrated all-chemical refinery operation. The return on investment after taxes is estimated at 16% for an all-motor fuel operation of the same magnitude. Little additional investment is required to produce an all-fuel refinery operation with pipeline gas as a major coproduct. The entire light oil fraction in this case is completely reformed to make automotive fuel. The comparison of the light oil crude is estimated as paraffinics 13 vol %, naphthenes 75 vol %, and aromatics 12 vol %. This gives for the production of the motor fuel pool an optional automotive fuel of better than 92 RON.

UT PROCESS FOR CARBONIZATION OF TEXAS LIGNITE

Introduction

Low temperature carbonization of lignite may be described as nondestructive distillation of coal. All coals of bituminous rank or lower will decompose upon heating to form primary tar, char and gas. The yields of these respective carbonization products will depend on the petrographic properties of the coal and the conditions under which carbonization takes place. Workers at the Bureau of

Mines at Denver showed that primary tar yields vary almost linearly with the heat content of the volatile matter in the coal (19).

Another important coal characteristic is its tendency to cake when heated. Most of the technical problems associated with coal carbonization involve caking, or agglomeration of coal in retorts upon heating. Because of its low caking tendencies and high volatile matter content, lignite, especially that found in parts of Texas appears to be the most attractive feedstock for a low temperature carbonization process.

The conditions of carbonization are the second group of factors influencing product yields. Investigators at Massachusetts Institute of Technology working on rapid devolatilization of pulverized coal showed, however, that decomposition of lignite is independent of pressure, heating rate, and particle size (20). Devolatilization depends only on temperature and residence time. This conclusion is supported, for the most part, by USBM data.

Tests have been conducted in both fixed bed assay units and entrained bed pilot plants to identify the optimum operating temperature for carbonization. In both instances, there was a dramatic break in the yield of primary tar versus temperature curve at approximately 850°F since the principal effect is the conversion of char into gas of decreasing heating value.

During the same study, it was shown that a particle mean residence time of 10 to 15 minutes was optimum when operating the carbonizer at 900°F. If the residence time is less than 10 minutes, the full tar yield is not realized. Operation of the retort with solids residence times greater than 15 minutes does not appreciably increase tar yield.

It is also important that sufficient volatile matter be present in the char to insure ready light-up and combustion. For pulverized coal, this corresponds to 15%, or more. Operation of the carbonizer retort in the above fashion will reduce the volatile matter content of the char to approximately 25%, well within the desired range.

Fluidized beds have been chosen as the mode of operation in the carbonizer retorts for several important reasons. Fluidized beds offer the advantages of continuous, automatically controlled processing, high unit capacities suitable for large-scale operations, high heat and mass transfer rates between gases and solid particles, uniform bed temperatures, and low equipment and processing costs. The main drawbacks of fluidized beds for this application are the nonuniform residence times for solids in the reactor and erosion of pipes and vessels from abrasion by entrained solids. These disadvantages can be minimized, however, by careful engineering design. Fluidized beds have been used successfully for several years to roast sulfide ores and reduce iron oxide under similar if not more severe operating conditions (21).

Low temperature char, the main product of carbonization, has been used extensively in Great Britain as a domestic fuel but has found little application for industrial firing. This is due in large measure to the lack of good outlets for by-products tar (22). Several workers have investigated the suitability of char as a boiler fuel, however, and found it to be quite attractive (23)(24). In particular, Phillips showed that low temperature chars with 10 to 20% volatile matter exhibit ignition characteristics similar to those of low-volatile bituminous coals. Burning test results indicate that low temperature chars will burn in commercial furnaces at rates similar to fast-burning coals.

If low temperature char is to be used to fire boilers for electric power generation, emission of particulates, SO_2 and NO_x must be controlled to meet Federal standards. In the conceptual design, venturi scrubbers and limestone-slurry absorbers are proposed to limit particulate emissions to 0.10 lb/MM Btu and SO_2 emission to 1.2 lb/MM Btu. Recombination of the scrubbed flue gases with hot carbonizer flue gases will raise the temperature of the combined stream to approximately 290°F, sufficient for plume bouyancy and well above the dew-point of the stack gas.

A study was carried out at the University of Texas at Austin into the possibilities of carbonization as a means of utilizing some of the 100 billion tons of lignite in Texas (25).

Process Description

A conceptual design of the UT Process as based on 500 MW gross power generation has been formulated by workers at the University of Texas.

The UT Process operates on run-of-mine lignite transported from the open mine to an on-site storage pile and then to the crusher house. Here lignite is reduced to an average particle diameter of approximately 900 microns. Crushed lignite is stored temporarily in bunkers before being moved by screw conveyers to fluidized-bed driers.

In the driers, the moisture content of the lignite is reduced from approximately 35 to 5 wt % as the lignite is heated to near 250°F. Heat required from drying comes in the form of hot gases generated by combustion of lignite fines carried over from the beds. Fines are collected by cyclones and then burned in a small cyclone furnace. Hot combustion gases, at approximately 2100°F, are used as the fluidizing medium, drying and heating the raw lignite. After passing through cyclones, these gases are combined with the boiler area flue gases for desulfurization and fine particulate removal. Three sets of fluidized-bed driers and cyclones are required for a plant generating 500 MW of electric power.

Dried lignite is directed to a small surge bunker and then conveyed by air to the carbonizer retorts. Each of the six retorts consists of a fluidized bed surrounded by an outer shell. Hot flue gas is circulated in the annular space between the two shells to supply part of the heat of carbonization. The remaining heat duty is supplied by combustion of a portion of the char in the fluidized bed. Air for combustion and lignite transport is compressed to 95 psia in an air compressor and then preheated to 700°F. The preheat is accomplished in two steps: air is heated to 410°F by compression to 80 psig; the remaining preheat is supplied by direct gas-to-gas heat exchange with make gas from the carbonizers in gas recuperator.

The tar vapors and gas formed upon carbonization at 900°F exit the top of the retorts at 950°F, pass through the gas recuperator, and are sent to high-efficiency cyclones for particulate removal. The cleaned vapors, at 820°F, are directed to the by-products processing area while the captured fines are recombined with the main char stream that leaves the carbonizers.

Tar vapors in the carbonizer off-gas stream are condensed in an air-cooled heat exchanger and removed from the gas stream by a vapor-liquid separator and an electrostatic precipitator. Tar is arbitrarily defined as the fraction of the hydrocarbon stream that boils below 190°F. The fraction that boils above 190°F is termed light oil. Light oil is recovered from the off-gas stream in a second train of cooler, vapor-liquid separator, and electrostatic precipitator. Final oil recovery takes place in an absorption column that uses tar as a wash oil. The oil which is collected before the absorber is decanted to remove water and mixed with the tar "rich" stream leaving the absorber.

This combined stream is pumped to the distillation section where it is separated into distillate and pitch. A flash still operating at 40 mm Hg and 475°F is used to effect the separation. Pitch from the flash still is sold as a by-product.

Tar acids in the distillate stream are separated by caustic extraction. Distillate is mixed with a 10% caustic (NaOH) solution to form sodium salts, or phenolates. Phenolates are converted to phenols and sodium carbonate in a springing tower by reaction with carbon dioxide from the char combustion section. Phenols, or tar acids, are decanted and pumped to the tar acid processing area. The sodium carbonate (Na_2CO_3) solution is mixed with lime (CaO) in an agitated tank to regenerate caustic along with precipitated limestone ($CaCO_3$). Limestone is removed by a rotary vacuum filter and then used in the SO_2 clean-up section of the power plant. This is one important example of integration in the UT Process.

The tar acids consist of phenols, cresols, xylenols and higher homologues. Acids boiling below 446°F are extremely valuable while those boiling above 446°F are not so valuable. In the tar acids processing section, low boiling acids (cresol and cresate) are separated from the high boiling fraction in a three-stage fractionating tower. The bottoms product is pumped to a fixed bed catalytic reactor where cracking takes place to produce more cresylics and a high boiling neutral oil which is highly aromatic. The high boiling fraction is separated in a second fractionating column and mixed with the residue from neutral oils processing to make a marketable aromatic feedstock. The top product from the second fractionation, containing cresylics and some low boiling neutral oils, is recycled as feed to the tar acid extraction section.

Tar acids represent only about 27% of the combined tar/light oil stream. The residual material that leaves the tar acids extraction section consists of neutral oils (67%) and tar bases (4%). The bases, primarily pyridine and higher homologues, are the most valuable chemicals, on a weight basis, present in the parent lignite. They are separated from the neutral oils by extraction with sulfuric acid and ammonia in succession. After extraction, the bases are fractionated into refined pyridine and technical grade pyridine. No credit is taken for the high boiling residue.

The material remaining after extraction of the acids and bases is a wide boiling range mixed residue of virtually every neutral hydrocarbon. Straight chained paraffins and olefins are extracted by urea adduction. The resulting material has proven to be a valuable feedstock for the production of biodegradable detergents. The residual material, when combined with the neutral oil stream from the tar acid processing section is a largely aromatic filtrate equally suitable for the production of maleic and phthalic anhydrides, or jet fuel.

The tar and light oil products of carbonization represents about 10 wt % of the original lignite feed, on a moisture- and ash-free basis. Almost 70 wt % of the original feed is recovered as char. The balance is made up by gas and water vapors.

Hot char is taken from the carbonizers and directed to pulverizers where the particle size is reduced so that 70% will pass a 200 mesh screen. This is the standard grind required for pulverized fuel combustion. It is hoped that continuous operation will allow direct firing of hot, pulverized char, thereby increasing the overall thermal efficiency of the process. Thirty days char storage facilities should be provided, however, to allow for load swings and carbonizer down-time.

The char is combusted in a pf furnace using 15% excess air. Equipment configurations for the furnace, superheater, economizer, and air heater are the same as those used in standard practice. The boiler generates 3,600,000 lb/hr of steam at 3,650 psi with superheat and reheat temperatures of 1000°F. This steam is expanded through a standard steam turbine to generate 500 MW of electric power.

Flue gases from the boiler area, as well as those from the lignite drier are directed to a limestone-slurry scrubbing process for cleaning. Fly ash is removed in a particulate venturi scrubber. A two-stage mobile bed scrubber is specified for sulfur dioxide removal on new power plants. After scrubbing, the flue gases are reheated by combination with carbonizer flue gas and then vented to the atmosphere.

An additional option for the UT Process is to replace the pulverized fuel boilers described above with gasifiers such as those built by Lurgi and Koppers-Totzek. The raw product gas can be used in a combined gas turbine-steam turbine power cycle for electric power generation.

Profitability of UT Process

The flow sheet was calculated assuming a power output of 800 MW. This set of conditions was taken as the base study for profitability calculations. Details of the feeds, product streams, operating conditions, and equipment specifications are given elsewhere (26)(27). To assign real values to the study, it was assumed that the project would begin in 1978 and the plant would begin operating on July 1, 1983.

The direct cost of equipment includes the purchased equipment costs for each unit, plus the cost of installation, foundations, supports, platforms, piping, instrumentation, electrical work, and painting. These items were estimated as percentages of the purchased equipment cost.

A summary of capital costs is shown in Table 7.10, assuming a processing rate of 3.25 million tons/yr of lignite. Line 8 gives the cost of the plant based upon 1975 purchased equipment costs. These values were accelerated to the start of construction using a 7.5% escalation rate. Then, assuming uniform cash flow throughout the two-year construction period, the cost of escalation during construction was determined. This gives the total depreciable cost of the plant. In order to avoid clouding the analysis, it is assumed here that the project is to be financed entirely with equity. However, the effect of debt financing was examined in the study and is available (27).

The plant is assumed to have the life expectancy of a typical petroleum facility which is fifteen years. Since no credit is taken for salvage value, depreciation is $3,608,000 per year on a straight line basis.

Table 7.10: Summary of Fixed Capital Costs (September 1975 Dollars)

Plant Section	Direct Cost	Indirect Cost	Total Fixed Capital
(1) Carbonization	6,570,000	2,630,000	9,200,000
(2) Tar and light oil separation	6,060,000	2,430,000	8,490,000
(3) Tar distillation	545,000	218,000	763,000
(4) Tar acids extraction	1,450,000	580,000	2,030,000
(5) Tar acids processing	2,420,000	968,000	3,388,000
(6) Tar bases extraction	1,620,000	650,000	2,270,000
(7) N-Hydrocarbon removal	3,630,000	1,450,000	5,080,000
(8) General facilities	1,390,000	560,000	1,950,000
Total (1975)	23,685,000	9,486,000	33,171,000
Escalated to start of construction at 7.5%/yr	35,933,000	14,373,000	50,306,000
Escalation during construction			3,808,000
Total Cost of Capital 1983			54,114,000
Depreciation per year			3,608,000

Source: *Energy Communications*

Operating Costs

Table 7.11 is a summary of the operating costs of the base study. The lignite price is the calculated cost of mining lignite in 1083, deflated to 1078 at 6.5% per year. All unit costs were adjusted to July 1, 1978, since that is when the UT Process hypothetically begins operations. This was done by escalating the 1978 prices at the rates shown in Table 7.12 which is compiled from published data and correspondence with industrial concerns. The following reasoning is relevant:

(1) Industrial labor escalation peaked out at 10% in 1974 in the chemical industry. In 1975 it was slightly lower and in 1976 fell to around 8%. Nor is it thought likely to go higher than this any time soon.

(2) Process materials are "energy-intensive" and since energy is at a premium these days, such items always "lead labor by a couple of points."

(3) Lignite is the only raw material for this process and it is mined on-site. Mining is a labor-intensive activity and therefore mining escalation rates should resemble those of labor. The energy shortage will cause a productivity increase which will make escalation rates of around 6.5%, i.e., about two points lower than labor.

(4) All utilities are supplied on-site and are therefore subject only to the escalation pressures of rising lignite costs. Therefore, 6% is used for these items also.

(5) Maintenance is labor-intensive, therefore 8% is used.

(6) Product Values—The products produced by the UT Process are highly energy-intensive and justifiably could be escalated as high as 10% per year. As the energy situation worsens, the free market value of these products, will grow at least this fast. However, in order to remain conservative, a safe 6.5% is used. Using this value, revenues are produced as shown in Table 7.13.

Table 7.11: Operating Costs of the UT Process (1978 Dollars)

Item	Usage/Year	1-1-78 Price	1-1-84 Price
Raw materials lignite	3.25×10^6 tons	3.24/ton	4.73/ton
Process materials*			
Lime	9,739 tons	42.25/ton	74.83/ton
Caustic	997 tons	295.06/ton	522.00/ton
H_2SO_4	4,284 tons	64.72/ton	114.66/ton
Ammonia	1,557 tons	253.80/ton	449.62/ton
Benzene	22,231 gal	1.02/gal	1.81/gal
Urea	1,482 tons	222.00/ton	393.29/ton
MIBK**	555 tons	634.53/ton	1,124.11/ton
Utilities			
Electricity	1.85×10^8 kWh	0.035/kWh	0.050/kWh
Cooling water	8.17×10^8 gal	0.046/1,000 gal	0.065/1,000 gal
Steam	7.96×10^8 lb	2.72/1,000 lb	3.86/1,000 lb
Other variables			
Sludge disposal	27,226 tons	2.41/ton	4.27/ton
Process water	1.7×10^7 gal	0.143/1,000 gal	0.218/1,000 gal
Other items			
Labor		7% of fixed capital	
Maintenance		5% of fixed capital	
Operating supplies		0.5% of fixed capital	
General plant overhead			
0.05 (labor) + 2 (operating supplies)		1.35% of fixed capital	
Sales expense		5% of manufacturing cost	
Catalyst		40% per year of initial catalyst investment	
Taxes and insurance		2% of fixed capital	

*From *Chemical Marketing Reporter.*
**Methyl isobutyl ketone for n-hydrocarbon removal process.

Source: *Energy Communications*

Table 7.12: Escalation Rates Assumed in Study

Construction, %	7.5
Product values, %	6.5
Raw materials, %	6.5
Process material costs, %	10.0
Utilities, %	6.0
Labor, %	8.0
Maintenance, %	8.0
Other expenses, %	6.5

Source: *Energy Communications*

Table 7.13: UT Process Revenues

Item	Production/Year	1-1-78 Price	1-1-84 Price
Cresylic acid	15,600 tons	\$0.520/lb	\$0.760/lb
2° Pyridine	1,653 tons	\$1.264/lb	\$1.844/lb
Technical pyridine	551 tons	\$0.400/lb	\$0.584/lb
N-hydrocarbons	18,755 tons	\$0.048/lb	\$0.070/lb
Aromatics	80,789 tons	\$0.176/gal	\$0.257/gal
Char	1.669 x 10^6 tons	\$5.680/ton	\$8.290/ton
Pitch	77,938.2 tons	\$5.680/ton	\$8.290/ton

Source: *Energy Communications*

On the basis of this information a profitability evaluation was made. Figures for the year 1978 are included in some of the tables in order to orient the reader. Depreciation is based upon instantaneous assembly of the plant on July 1, 1978. 1983 is the first year of plant operations; 1997 is the last. An operating year runs from July 1 to June 30 of the following year.

The above analysis gives a 1978 discounted rate of return of 13.5% after taxes.

Conclusions

Studies at the University of Texas have shown that the UT Process is both technically and economically feasible. Technical problems associated with the process are minimized because of heavy borrowing from existing, proven technology. This is especially true in the coal crushing and drying sections and the char combustion section where most of the equipment is "off-the-shelf" hardware. The carbonizer retorts are simple in design which should reduce operating problems.

Sound engineering design will be required, however to overcome anticipated problems in the solid handling equipment and erosion and corrosion in the vapor take-off line from the carbonizer. The unit operation required in the tar separation and processing sections are standard to the coke oven industry.

Based for the most part on sound processing technology, the UT Process offers a novel and viable method to make full and efficient utilization of lignite as an energy source. The process has as its primary objective to strip off the tars and light oils present as volatile matter in raw lignite so that these can be separated and processed to valuable chemical feedstocks. At the same time, lignite is upgraded from its raw form to a high carbon-content, highly reactive char that can serve as an acceptable fuel source either for a conventional generating plant or an integrated gasification/combined cycle power system.

REFERENCES

(1) Fischer, F. and Schrader, H., *Hydrogenation of Coal with Carbon Monoxide,* Vol 2, 257-261 (1921).

(2) Appell, H.R. and Wender, I., "The Hydrogenation of Coal with Carbon Monoxide and Water," Div. of Fuel Chem. Preprints, American Chemical Society, Vol 12, No. 3, 220-222 (1968).

(3) Appell, H.R., Miller, R.D. and Wender, I., "On the Mechanism of Lignite Liquefaction with Carbon Monoxide and Water," presented before the Div. of Fuel Chem., 163rd National Meeting, American Chemical Society (April 10-14, 1972).
(4) Appell, H.R., Moroni, E.C. and Miller, R.D., "CO Steam Liquefaction of Lignite," Div. of Fuel Chem., preprints, American Chemical Society, Vol 20, No. 1 (58-65).
(5) Appell, H.R., Wender, I., and Miller, R.D., *Liquefaction of Lignite with Carbon Monoxide and Water,* Bureau of Mines IC 8543, 32-39 (1972).
(6) York, W.J., "Reduction of Subbituminous Coal and Lignite Using Carbon Monoxide," Doctoral dissertation, Montana State University (1972).
(7) Fiske, T.A., "Reduction of Subbituminous Coal Using Carbon Monoxide," Masters thesis, Montana State University (1972).
(8) Nguyen, D., "Reduction of Subbituminous Coal with Carbon Monoxide and Water," Doctoral dissertation, Montana State University (1973).
(9) Bodner, S.S., "Effect of Catalysts on the Reduction of Subbituminous Coal by the CO-Steam Process," Masters thesis, Montana State University (1977).
(10) Del Bel, E., Friedman, S., Yavorsky, P.M. and Wender, I., "The Liquefaction of Lignite by the CO-Steam Process," presented at the A.I.Ch.E. National Meeting, Houston, TX (March 16-20, 1975).
(11) Knudson, C.L., Schiller, J.E. and Ruud, A.L., "Temperature Effects on Coal Liquefaction; Rates of Depolymerization and Product Quality as Determined by Gel Permeation Chromatography," presented before the Div. of Fuel Chem., American Chemical Society, National Meeting, Chicago, IL (Aug. 1977).
(12) Schiller, J.E. and Mathiason, D.N., "A Simple Separation Method for Coal Derived Solids and Heavy Liquids," *Analytical Chemistry,* in press.
(13) Schiller, J.E., "Composition of Coal Liquefaction Products," *Hydrocarbon Processing,* No. 1, 147-152 (1977).
(14) Schiller, J.E., "Characterization of Coal Derived Liquids," presented at the Great Lakes Regional Meeting, American Chemical Society, Stevens Point, WI (June 1977).
(15) Schiller, J.E., Farnum, B.W. and Sondreal, E.A., "Viscosity of Coal Liquids—The Effect of Character and Content of the Non-Distillable Portion," presented before the Div. of Fuel Chem., American Chemical Society, National Meeting, Chicago, IL (Aug. 1977).
(16) Souby, A.M., Severson, D.E. and Kube, W.R., "Project Lignite: Convenience Fuels from Northern Great Plains Province Lignite," *N. Dakota Acd. of Sci.,* 28, Part II, 50-59 (Sept. 1976).
(17) Scarrah, W.P. and Dillon, R.R., "Startup Solvent Selection for the Liquefaction of Lignite," *Ind. Eng. Chem Process Des. Dev.,* Vol 15, No. 1, 122-129 (1976).
(18) Nowacki, P., *Coal Liquefaction Processes,* Noyes Corporation, Park Ridge, NJ (1979).
(19) Perry, V.F., Landers, W.S., Wagner, E.O., Goodman, J.B. and Lammers, G.C., *Drying and Carbonizing Fine Coal in the Entrained and Fluidized State,* USBM Report of Investigation 4954 (1953).
(20) Anthony, D.B., Howard, J.B., Hottel, H.C. and Meisner, H.P., "Rapid Devolatilization of Pulverized Coal, " presented at the Fifteenth International symposium on Combustion, Tokyo (1974).
(21) Kunii, D. and Levenspiel, O., *Fluidization Engineering,* John Wiley and Sons, Inc., NY (1969).
(22) Lowry, H.H. (Editor), *Chemistry of Coal Utilization: Supplementary Volume,* John Wiley and Sons, Inc., NY (1963).
(23) Graham, J., "Combustion Optimization," *Electrical World,* 18B, 43-58 (June 15, 1976).
(24) Aul, E.F., Jr., "Conceptual Design and Technical Feasibility Study of the UT Process for Low-Temperature Carbonization of Texas Lignite," M.S. thesis, The University of Texas at Austin, Austin, TX (1977).
(25) Kaiser, W.R., "Texas Lignite: Near Surface and Deep Basin Resources," University of Texas at Austin, Bureau of Economic Geology, Department Investigation No. 79 (October 1974).
(26) Aul, E.F., "Conceptual Design and Technical Feasibility of Low Temperature Carbonization of Texas Lignite," M.S. thesis, University of Texas (May 1977).
(27) Massey, N.A., "The Economic Feasibility of Carbonization of Texas Lignite," M.S. thesis, University of Texas (May 1977).

COAL LIQUEFACTION PROCESSES 1979

by Perry Nowacki

Chemical Technology Review No. 131
Energy Technology Review No. 45

With the precarious position of petroleum supplies in the world market and the unfavorable balance of payments caused by overdependence on foreign sources, it may be expedient to exploit coal further as a source of liquid fuels. In the United States coal is the most abundant natural energy resource. Through its conversion to liquid fuels, it could supply a portion of the energy required for transportation, heating, etc.

Coal can furnish a complete range of liquid fuels, as well as chemical feedstocks. To achieve maximum yields, however, judicious adjustment of temperature, pressure, time and catalyst, as elaborated in this book, is essential. The text reviews the histories of important domestic and foreign coal liquefaction processes—pyrolysis, solvent extraction, and catalytic and indirect liquefaction. It details the chemistry of the processes and their inherent problems, also discussing economics, developmental status, and environmental impacts.

Chapter headings and **examples of some** important subtitles follow below:

1. **INTRODUCTION**
2. **COAL RESOURCES**
3. **COAL LIQUEFACTION CHEMISTRY**
4. **OVERVIEW OF PROCESSES**
5. **PYROLYTIC PROCESSES**
   Char Oil Energy Development (COED)
   Clean Coke
   TOSCOAL
   Occidental Flash Pyrolysis
   Coalcon Hydrocarbonization
   Oak Ridge Hydrocarbonization
   Flash Hydropyrolysis of Lignite
   Flash Hydrogenation
   Lurgi-Ruhrgas Flash Carbonization
6. **SOLVENT EXTRACTION**
   Solvent-Refined Coal (SRC) Process
   Pilot Plants—Ft. Lewis & Wilsonville
   Consol Synthetic Fuel Process
   Cresap Pilot Plant
   Exxon Donor Solvent Process
   CO-Steam Process
   Supercritical Gas Solvent Extraction
   UOP Solvent Extraction
7. **CATALYTIC LIQUEFACTION**
   H-Coal Process
   Trenton, N.J. Pilot Plant
   Synthoil Process
   Clean Fuel from Coal Process
   Zinc Halide Coal Liquefaction
   Dow Coal Liquefaction
8. **INDIRECT LIQUEFACTION**
   Fischer-Tropsch Process
   SASOL Commercial Plant, South Africa
   Mobil M-Gasoline Process
9. **COAL LIQUIDS FROM H-COAL**
   Analysis and Properties of Products
   Distillate Separation & Characterization
10. **SOLVENT-REFINED EXTRACT—COMMERCIAL, EXPANDED-BED HYDROPROCESS**
11. **ECONOMICS OF SRC PROCESS**
   Reclaiming, Preparing, Liquefying Coal
   Hydrogenation of Fuel Oil and Naphtha
   Stack Gas Desulfurization
   Hydrogen Manufacture
   Capital Investment and Costs
12. **ENGINEERING ASSESSMENT OF SYNTHOIL PROCESS**
   Feed Slurry Pumping and Heating
   Reactor Design
   Removal of Spent Catalyst
   Low-Temperature Carbonization
13. **FISCHER-TROPSCH DESIGN & COST**
   Recovery of Heat, Liquids, Sulfur
   Gasification and Methanation
   Water Systems
14. **EQUIPMENT AND MATERIALS**
   Pumps, Heat Exchangers, Fired Heaters
   Vessels and Pipe Systems
   Instrumentation and Metallurgy
15. **ENVIRONMENTAL, HEALTH AND SAFETY ISSUES**
   Solid Waste Management
   Hazards in Product Transfer and Storage
   Federal Legislation
16. **COMPARISON OF PROCESSES**
   Technical and Economic Aspects
   Probability of Commercial Success

ISBN 0-8155-0756-9 **339 pages**